PAL 3.1

practice
anatomy
lab™

Lab Guide

PAL 3.1 | practice anatomy lab™

Lab Guide

Ruth E. Heisler
University of Colorado at Boulder

Nora Hebert
Red Rocks Community College

with

Jett Chinn
Cañada College and College of Marin

Karen M. Krabbenhoft
University of Wisconsin – Madison,
School of Medicine & Public Health

Olga Malakhova
University of Florida College of Medicine, Gainesville

PEARSON

Boston Columbus Indianapolis New York San Francisco Upper Saddle River
Amsterdam Cape Town Dubai London Madrid Milan Munich Paris Montréal Toronto
Delhi Mexico City São Paulo Sydney Hong Kong Seoul Singapore Taipei Tokyo

Editor-in-Chief: Serina Beauparlant
Project Editor: Nicole Graziano
Senior Development Manager: Barbara Yien
Assistant Editor: Lisa Damerel
Senior Instructional Designer: Sarah Young-Dualan
Senior Managing Editor: Debbie Cogan
Production Project Manager: Caroline Ayres
Production Editor: Mary Tindle
Copyeditor: Mike Rossa
Compositor: S4Carlisle Publishing Services

Interior and Cover Designer: Tammy Newnam
Image Lead: Donna Kalal
Art House: Precision Graphics, Kristina Seymour;
    Pearson Digital Pub, Cory Skidds
Photo Researcher: Bill Smith Group
Senior Manufacturing Buyer: Stacey Weinberger
Marketing Manager: Derek Perrigo
Cover Photo Credits: Karen Krabbenhoft, Pearson
    Science; Larry DeLay, Pearson Science;
    Nina Zanetti, Pearson Science

23 2020

www.pearsonhighered.com

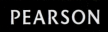

ISBN 10: 0-321-84025-9
ISBN 13: 978-0-321-84025-7
(Student edition standalone)

ISBN 10: 0-321-85767-4
ISBN 13: 978-0-321-85767-5
(Student edition with PAL™ 3.1 DVD)

Preface

Practice Anatomy Lab™ 3.1 software allows for an interactive laboratory experience, regardless of whether a student has direct access to a cadaver, models, or a microscope. The purposes of the companion Lab Guide are to provide guidance and to enhance the learning experience through a series of labeling activities and quizzes that directly correlate to PAL 3.1. In addition, there are functional questions and activities that take students beyond PAL and encourage application of the anatomy to more physiological and clinically relevant situations. The Lab Guide also provides instructors with a lab manual of activities that they can assign during lab or outside lab.

The Lab Guide explores anatomical structures using the standard undergraduate systems approach. Within each system, students will first explore the gross anatomy and conclude with appropriate histological images. The Lab Guide reflects the content found in the Human Cadaver, Anatomical Models, and Histology modules in PAL 3.1.

Overview of PAL 3.1™ Lab Guide

The chapters are organized by body system and correspond to the 11 body systems in PAL 3.1. Each body system includes the following:

- **Self-review** activities ask the student to find, identify, and label the structures most commonly taught in an undergraduate anatomy lab course. Some chapters (Ch. 2 Skeletal System and Ch. 3 Muscular System) include exercises that require students to interact with the 3-D animations and bone rotations in PAL 3.1. For these exercises, students are directed to specific animations in PAL 3.1. After viewing the clips, students are asked to complete tables that detail the origins, insertions, actions, and innervations of select muscles.

- **Quiz** follows each Self Review and includes two levels of questions. The first level, "Check Your Understanding," consists of short-answer, fill-in-the-blank, and matching questions focusing on structure and basic function of key anatomical structures. Some of these questions are more comprehensive, requiring students to utilize their textbook and other class resources. The second level, "Apply What You Learned," consists of more challenging short-answer questions that force students to apply, analyze, and sometimes synthesize their knowledge of anatomy while considering a real-world scenario.

- **Lab Practical** concludes each chapter, which simulates a real lab exam with fill-in-the-blank questions. Here, students can test their knowledge of location, spelling, and function of the most important anatomical structures while studying color images from PAL 3.1.

PAL™ 3.1 Software

PAL 3.1 is an indispensable virtual anatomy study and practice tool that gives students 24/7 access to the most widely used lab specimens including human cadaver, anatomical models, histology, cat, and fetal pig.

PAL 3.1 is available as a DVD, an app for both iPad and Android tablets, and a website in the Study Area of MasteringA&P®.

Key features of PAL 3.1 include:

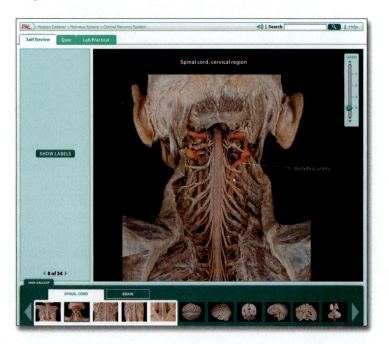

- **New interactive cadaver** that allows students to peel back layers of the human cadaver and view hundreds of brand new dissection photographs specially commissioned for version 3.1. Carefully prepared dissections show nerves, veins, and arteries across body systems.

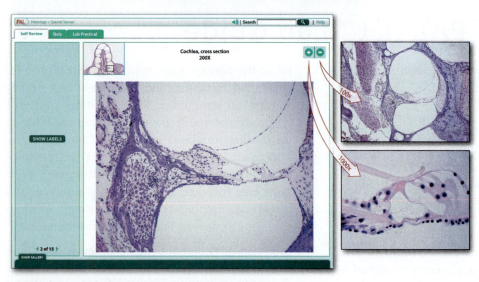

- **New interactive histology** module that allows students to view the same tissue slide at varying magnifications, thereby helping them identify structures and their characteristics.

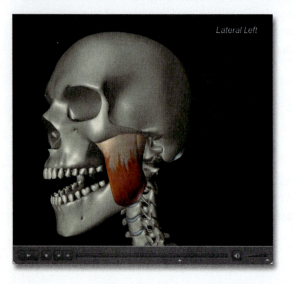

- **3-D anatomy animations** of origins, insertions, actions, and innervations for over 60 muscles. PAL 3.1 also includes over 50 anatomy animations of group muscle actions and joints.

- **New question randomization feature** that gives students more opportunities to practice. Each time the student retakes a practice quiz or lab practical a new set of questions is generated.

PAL™ 3.1 Instructor Resources

The following instructor resources are available for PAL 3.1:

- **Instructor's Resource DVD for PAL 3.1** provides instructors access to all images in PAL in PowerPoint® and JPEG formats. PowerPoint slides include images with editable labels and leader lines, as well as embedded links to relevant 3-D anatomy animations and bone rotations.

- **PAL 3.1 Test Bank** includes more than 4,000 customizable multiple-choice quiz and fill-in-the-blank lab practical questions in TestGen® format. The Test Bank is available for download in the Instructor Resource Center and is also fully assignable in the MasteringA&P® Item Library.

- **PAL 3.1 Lab Guide Answer Key** is available only to instructors and can be downloaded at MasteringA&P Instructor Resources or at Pearson Instructor Resource Center at www.pearsonhighered.com/irc.

Packaging Options

Flexible packaging and ordering options are available:

- **The Practice Anatomy Lab 3.1 DVD** can be packaged with the Lab Guide. (ISBN 978-0-321-85767-5 / 0-321-85767-4)

- **The Lab Guide is also available without the PAL 3.1 DVD** for those students who already have access to PAL 3.1 through one of the following options:
  - MasteringA&P Study Area, which accompanies selected Pearson texts and lab manuals.
  - PAL 3.1 Mobile App, which is available for iPad and Android tablet devices and can be accessed using a MasteringA&P login and password.
  - A 12-month online subscription to PAL 3.1, which is available at www.practiceanatomylab.com.

- **The Lab Guide can also be customized through Pearson's Custom Group.** Plans for Activities for the Cat and Fetal Pig specimens in PAL 3.1 are under consideration and will possibly be available through Pearson Custom. Please consult with your Pearson representative for more information.

Acknowledgments

Education is our passion and the amazing team at Pearson has allowed us to reach more students than we ever thought possible. We particularly want to acknowledge our astute Editor-in-Chief Serina Beauparlant who is not only an insightful businesswoman but is also a true champion of her authors; Project Editor Nicole Graziano who has held our hands throughout this whole endeavor and skillfully managed our many angst-filled phone calls; and Senior Instructional Designer Sarah Young-Dualan who has guided us from the very beginning. These amazing individuals, along with the many talented reviewers, have made this the best product possible.

A special thank you to the following individuals who have also contributed to PAL:

Samuel Chen, *Moraine Valley Community College*
Larry DeLay, *Tri-Power Performance, Inc.*
Stephen W. Downing, *University of Minnesota, Medical School*
Lisa M. J. Lee, *The Ohio State University, College of Medicine*
Eksel Perez and Peter Westra, *San Francisco, CA*
Winston Charles Poulton, *University of Florida College of Medicine, Gainesville*
Leif Saul, *University of Colorado at Boulder*
Renn Sminkey, *Creative Digital Visions, LLC*
Michael J. Timmons, *Moraine Valley Community College*
Nina Zanetti, *Siena College*

We would like to thank the following academic reviewers for their valuable contributions to the Lab Guide:

Michele Alexandre, *Durham Technical College*

Claudie Biggers, *Amarillo College*

Patty Bostwick-Taylor, *Florence-Darlington Technical College*

Lance Brand, *Ball State University*

Betsy Brantley, *Valencia Community College*

Bertha Byrd, *Wayne State Community College*

John Capodilupo, *Grand Valley State University*

Marnie Chapman, *University of Alaska – Sitka*

Chet Cooper, *Odessa College*

Donna Crapanzano, *Stony Brook University*

Galen DeHay, *Tri-County Technical College*

Wendy Dusek, *Wisconsin Indianhead Technical College*

Christina Gan, *Highline Community College*

Edwin Gines-Candelaria, *Miami Dade College*

Leslie Hendon, *University of Alabama – Birmingham*

Melinda Hutton, *McNeese State University*

Kimberly Kerr, *Troy University*

Beth Kersten, *Manatee Community College*

Nancy Kincaid, *Troy University*

Brian Kuyatt, *Hillsborough Community College*

Laura Mastrangelo, *Hudson Valley Community College*

Julie Rosenheimer, *Texas Tech University*

Mohtashem Samsam, *University of Central Florida*

Miriam Satern, *Western Illinois University*

Steve Schneider, *South Texas College*

Patricia Wilhelm, *Community College of Rhode Island – Warwick*

James Yount, *Brevard Community College*

About the Authors

Professor Ruth Heisler
Ruth Heisler is a senior instructor in the Department of Integrative Physiology at the University of Colorado at Boulder where she teaches and coordinates several courses, including Human Anatomy, Comparative Vertebrate Anatomy, and Forensic Biology. Ruth received her B.S. in Biology from the University of Minnesota, and her M.A. in Biology from the University of Colorado. She has been an instructor at the University of Colorado for 14 years.

At the University of Colorado, Ruth has worked extensively with the Science Education Initiative to improve both the teaching and understanding of scientific material at the undergraduate level. In addition, she has been involved in academic outreach through workshops with the American Academy of Forensic Sciences and the Biological Sciences Initiative. She has been a consultant on projects with the Center for Human Simulation, working with the Visible Human Project. Additionally, she is an active member of the Human Anatomy and Physiology Society (HAPS), where she has presented workshops on the use of *Practice Anatomy Lab* as an assessment tool in the classroom.

Ruth has been deeply involved in the development of *Practice Anatomy Lab* for nearly six years, as coauthor of versions 2.0, 3.0 and 3.1. She is also author of a custom lab manual developed for a cadaver-based human anatomy lab. In her spare time, Ruth enjoys spending time in the mountains and traveling with her husband and two young sons.

Dr. Nora Hebert
Dr. Nora Hebert teaches undergraduate courses in Anatomy and Physiology at Red Rocks Community College near Denver, Colorado. Although most of her students are undergraduates, primarily interested in the allied health professions, Nora has also taught graduate-level Human Physiology for the College's Physician Assistant Program.

Nora received a Ph.D. in Endocrinology from the University of California at Berkeley. Although her research interests have primarily been in basic science, she became familiar with the research phase of drug development while doing a postdoctoral fellowship in the private sector. It was during this fellowship that Nora decided to pursue teaching as a profession.

Nora is an active faculty member at Red Rocks, serving on the faculty senate, the honors program committee, and the admissions and executive committees for the Physician Assistant Program. She is also part of the College's Campus Green Initiative.

Nora has consulted in the development of the Explorable Virtual Human with the Center for Human Simulation at the University of Colorado Health Sciences Center. She has also been involved with the Visible Human Dissector program, advising K–12 teachers and postsecondary instructors on how best to implement the Dissector in their classrooms.

Nora has been deeply involved in the development of *Practice Anatomy Lab* for nearly six years, as coauthor of versions 2.0, 3.0 and 3.1. She is also the author of over 60 A&P Flix animations covering muscle physiology, neurophysiology, and muscle origins, actions, insertions, and innervations.

Professor Jett Chinn
Jett Chinn is an instructor of Human Anatomy in the Science and Technology Division of Cañada College and also the Life and Earth Sciences Department at the College of Marin.

Mr. Chinn has over 20 years of experience teaching Human Anatomy at institutions including San Francisco State University, California College of Podiatric Medicine, Touro University College of Osteopathic Medicine, and UC San Francisco School of Medicine.

Mr. Chinn received a B.A. in general biology from San Francisco State University.

Dr. Karen Krabbenhoft
Dr. Karen Krabbenhoft is a senior lecturer in the Department of Neuroscience at the University of Wisconsin in Madison. During her 20-year career, Dr. Krabbenhoft's focus has been on teaching students at all levels of their educational process, including undergraduate, physician assistant, and medical students.

Dr. Krabbenhoft earned her Ph.D. in Anatomy from the University of Wisconsin.

Dr. Olga Malakhova
Dr. Olga Malakhova is an assistant scholar in the Department of Anatomy and Cell Biology at the University of Florida College of Medicine in Gainesville. She has been teaching first-, second-, and fourth-year medical students, as well as several Clinical Residency programs, at the University of Florida for the past 20 years.

Dr. Malakhova received her M.D. from Odessa Medical Institute in Ukraine, and her Ph.D. in Neuroscience from the Brain Research Institute of the Russian Academy of Medical Sciences in Moscow.

Contents

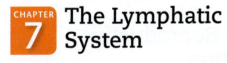

CHAPTER 7 The Lymphatic System

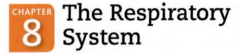

CHAPTER 8 The Respiratory System

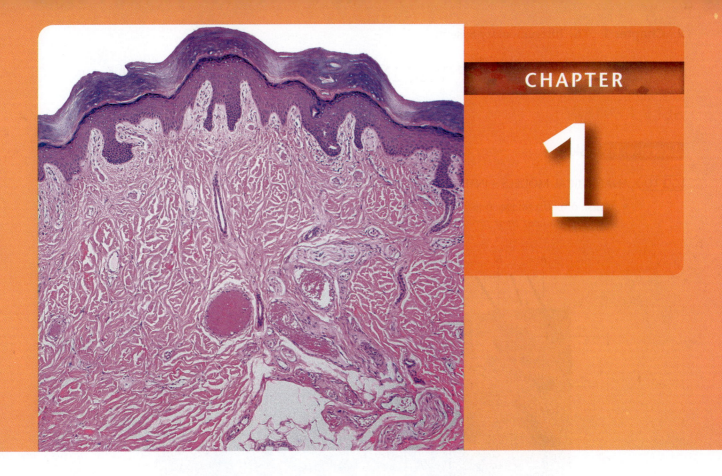

The Integumentary System

STUDENT OBJECTIVES

1. Identify the epidermis and dermis, and describe the structures found in each layer.
2. List in order (apical to basal) the layers of the epidermis and describe the characteristics of each layer.
3. Identify the sublayers of the dermis and describe the characteristics of each layer.
4. Identify which structures of the integument are derived from the epidermis.
5. Understand the important functions of the integumentary system.
6. Describe the location and functions of sebaceous glands, sweat glands, and hair.

GROSS ANATOMY OF THE INTEGUMENTARY SYSTEM

SELF REVIEW ⎤ Anatomical Models

Exercise 1.1 **Skin**

GO TO ⟩ ANATOMICAL MODELS > INTEGUMENTARY SYSTEM > SELF REVIEW > IMAGE 1

- *Mouse over the image to locate and label the structures indicated below. Click on the structures to hear their pronunciations.*

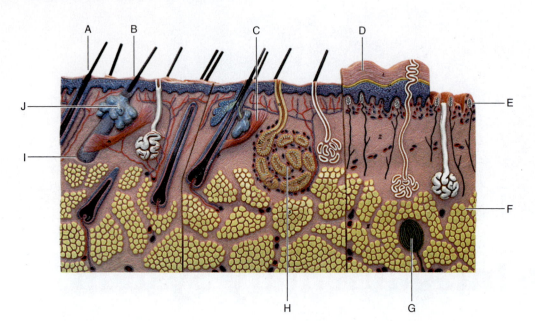

Copyright by SOMSO, 2010, www.somso.com

A. _____

B. _____

C. _____

D. (layer) _____

E. _____

F. (layer) _____

G. _____

H. _____

I. _____

J. _____

Exercise 1.2 Epidermis (Thick Skin)

GO TO ANATOMICAL MODELS > INTEGUMENTARY SYSTEM > SELF REVIEW > IMAGE 2

> • *Mouse over the image to locate and label the structures indicated below. Click on the structures to hear their pronunciations.*

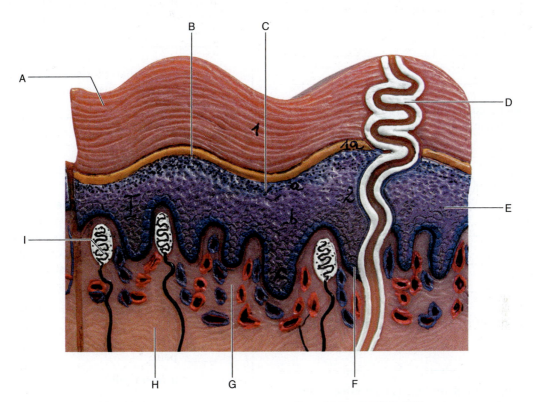

A. _____

B. _____

C. _____

D. _____

E. _____

F. _____

G. (layer) _____

H. (layer) _____

I. _____

QUIZ Anatomical Models

Answer the following questions using information from PAL 3.1 as well as other course materials including your textbook, lecture, and lab notes.

I. Check Your Understanding

1. A hair has two basic regions. Identify these regions.

   _____

   _____

2. Meissner's corpuscles are found in which of the two basic layers of the dermis?

   _____

3. Identify the five layers of the epidermis in thick skin.

   _____

   _____

   _____

4. What type of sweat gland is prominent in thick skin?

   _____

5. The wall of a hair follicle is composed of two sheaths. Identify these two sheaths.

   _____

   _____

BEYOND PAL 6. The superficial layers of the epidermis contain two chemicals that waterproof the skin and protect it from wear and tear. Identify these two chemicals. In which layer of the epidermis are these chemicals made?

   _____

   _____

   _____

7. One of the functions of human skin is to detect the movement of insects. What receptor is activated by crawling insects?

   _____

8. Describe the characteristics and the functions of sebaceous gland secretions.

   _____

   _____

   _____

9. Match the structures on the right to the location on the left. There may be more than one answer for each location on the left. However, there is only one correct combination when all the structures on the right are used.

   _____ hair follicle

   _____ dermis

   _____ palm

   _____ arrector pili

   _____ sebaceous gland

   _____ axilla

   _____ stratum granulosum

   a. Sebum

   b. Smooth muscle

   c. Apocrine sweat gland

   d. Collagen fibers

   e. Eccrine sweat gland

   f. Keratohyaline

   g. Epithelial root sheath

II. Apply What You Learned

1. List the four ways by which skin can be burned. Classify burns according to their severity or depth. Include details on symptoms, damage to specific structures, and the predicted healing time.

_____

_____

_____

_____

_____

_____

2. Metastatic melanoma is an especially deadly form of skin cancer. Which skin cell type rapidly divides to form a melanoma? The American Cancer Society advises the application of the ABCDE rule for recognizing melanoma. The acronym ABCDE stands for the diagnostic features characteristic of melanoma. Define these features. Why is melanoma highly metastatic?

_____

_____

_____

_____

_____

_____

_____

TISSUES OF THE INTEGUMENTARY SYSTEM

SELF REVIEW | Histology

Exercise 1.3 Thick Skin, Longitudinal Section 40x

GO TO Histology > Integumentary System > Self Review > Image 1

- _Mouse over the image to locate and label the structures indicated below. Click on the structures to hear their pronunciations._

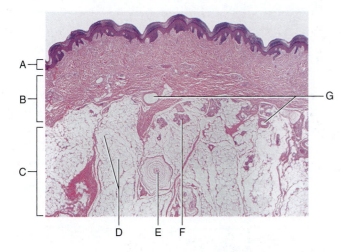

A. _____

B. _____

C. _____

D. _____

E. _____

F. _____

G. _____

Exercise 1.4 **Thick Skin, Dermis and Hypodermis, Longitudinal Section 100x**

GO TO HISTOLOGY > INTEGUMENTARY SYSTEM > SELF REVIEW > IMAGE 3

- *Mouse over the image to locate and label the structures indicated below. Click on the structures to hear their pronunciations.*

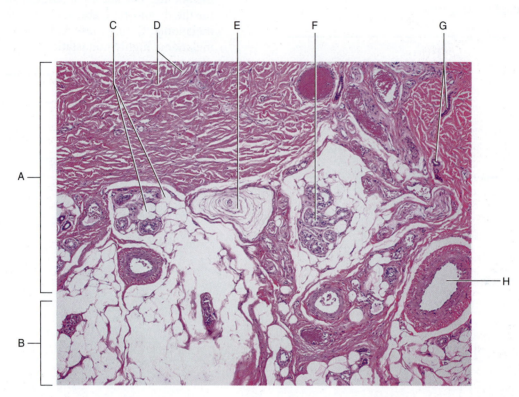

A. _____   E. _____

B. _____   F. _____

C. _____   G. _____

D. _____   H. _____

Exercise 1.5   **Thick Skin, Epidermis and Dermis, Longitudinal Section 400x**

GO TO HISTOLOGY > INTEGUMENTARY SYSTEM > SELF REVIEW > IMAGE 4

- *Mouse over the image to locate and label the structures indicated below. Click on the structures to hear their pronunciations.*

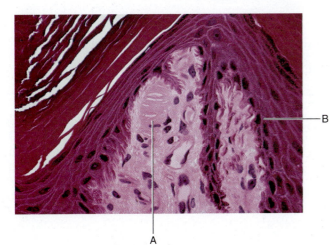

A. _____

B. _____

C. _____

D. _____

E. _____

F. _____

G. _____

H. _____

Exercise 1.6   **Thick Skin, Meissner's Corpuscle, Longitudinal Section 1000x**

GO TO HISTOLOGY > INTEGUMENTARY SYSTEM > SELF REVIEW > IMAGE 8

- *Mouse over the image to locate and label the structures indicated below. Click on the structures to hear their pronunciations.*

A. _____

B. _____

Exercise 1.7 Pacinian Corpuscle, Longitudinal Section 200x

GO TO HISTOLOGY > INTEGUMENTARY SYSTEM > SELF REVIEW > IMAGE 9

- *Mouse over the image to locate and label the structures indicated below. Click on the structures to hear their pronunciations.*

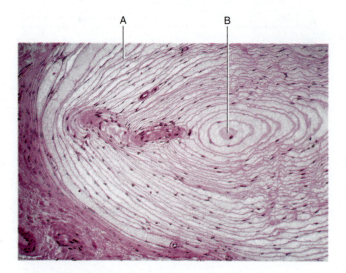

A. _____

B. _____

Exercise 1.8 Thin Skin, Hair Follicle, Longitudinal Section 100x

GO TO HISTOLOGY > INTEGUMENTARY SYSTEM > SELF REVIEW > IMAGE 11

- *Mouse over the image to locate and label the structures indicated below. Click on the structures to hear their pronunciations.*

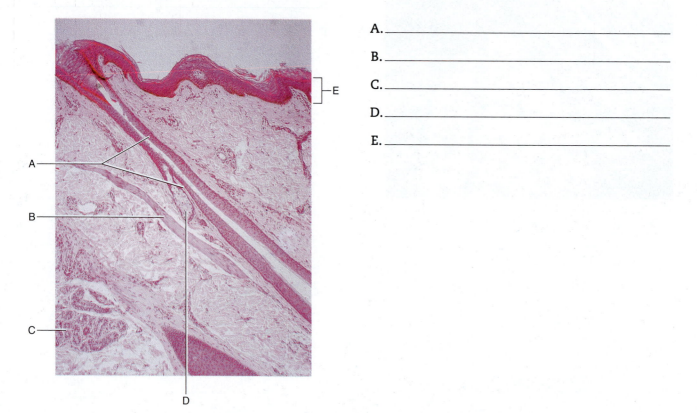

A. _____

B. _____

C. _____

D. _____

E. _____

Exercise 1.9 Thin Skin, Hair Follicle, Longitudinal Section 150x

GO TO > HISTOLOGY > INTEGUMENTARY SYSTEM > SELF REVIEW > IMAGE 12

- *Mouse over the image to locate and label the structures indicated below. Click on the structures to hear their pronunciations.*

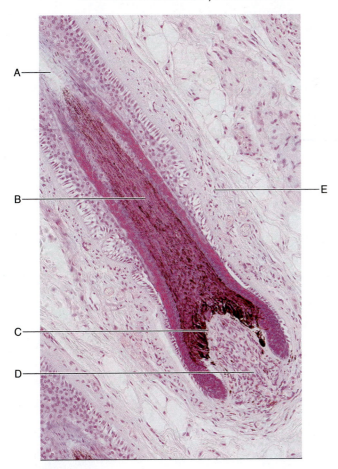

A. _____

B. _____

C. _____

D. _____

E. _____

Exercise 1.10 Thin Skin, Hair Follicle, Cross Section 200x

GO TO > HISTOLOGY > INTEGUMENTARY SYSTEM > SELF REVIEW > IMAGE 14

- *Mouse over the image to locate and label the structures indicated below. Click on the structures to hear their pronunciations.*

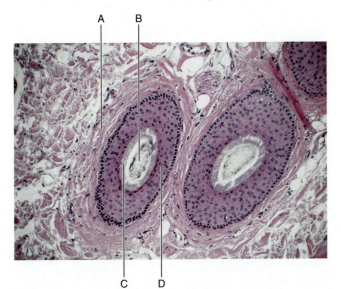

A. _____

B. _____

C. _____

D. _____

Exercise 1.11   **Thin Skin, Hair Follicle, Longitudinal Section 200x**

GO TO HISTOLOGY > INTEGUMENTARY SYSTEM > SELF REVIEW > IMAGE 15

- *Mouse over the image to locate and label the structures indicated below. Click on the structures to hear their pronunciations.*

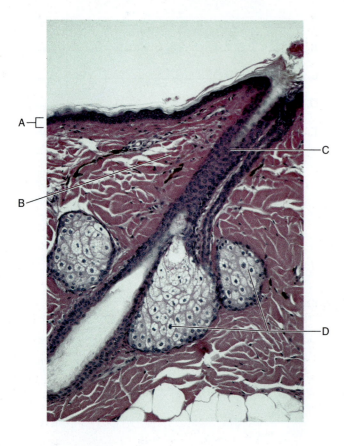

A._____

B._____

C._____

D._____

Exercise 1.12   **Thin Skin, Longitudinal Section 400x**

GO TO HISTOLOGY > INTEGUMENTARY SYSTEM > SELF REVIEW > IMAGE 17

- *Mouse over the image to locate and label the structures indicated below. Click on the structures to hear their pronunciations.*

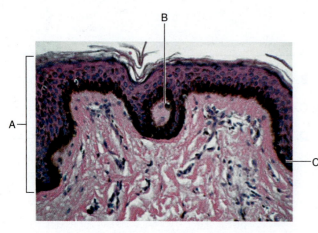

A. (epithelium) _____

B._____

C._____

QUIZ) Histology

Answer the following questions using information from PAL 3.1 as well as other course materials including your textbook, lecture, and lab notes.

I. Check Your Understanding

1. Is the dermis of the skin composed of dense regular or dense irregular connective tissue?

_____

2. Which layer(s) of the epidermis consist(s) of many layers of dead, keratinized cells?

_____

3. Which layer of the dermis interacts with the epidermis to form epidermal ridges?

_____

4. Which mechanoreceptor, consisting of a central nerve ending surrounded by concentric layers of connective tissue, deforms when deep pressure is applied to the skin?

_____

5. Using Self Review image 11 of 17 in Histology > Integumentary System, determine whether hair follicles are derived from the dermis or the epidermis.

_____

BEYOND PAL **6.** Which cells are the most abundant within the epidermis? Note that these cells are continuously removed from the apical surface of the epidermis. Therefore, they must be replaced by cell division. Which layer of the epidermis is responsible for this mitotic activity?

_____

_____

7. What vitamin precursor is formed within dermal blood vessels in response to UV light? Why is the mature form of this vitamin so critically important?

_____

_____

_____

_____

_____

_____

8. Order the layers of the epidermis, at right, starting with the apical layer (1) and ending with the basal layer (5).

1. _____   a. Stratum lucidum

2. _____   b. Stratum basale

3. _____   c. Stratum spinosum

4. _____   d. Stratum corneum

5. _____   e. Stratum granulosum

II. Apply What You Learned

1. On the one hand, UV radiation is necessary for the production of a key vitamin. On the other, too much UV radiation is harmful to the skin. Describe why overexposure to UV radiation is damaging.

   _____

   _____

   _____

   _____

   _____

   _____

   _____

2. Acne develops when the duct of a sebaceous gland becomes clogged with sebum (oil) and dead skin cells. Explain why acne appears mostly on the face, neck, chest, back, and shoulders—but not on the palm of the hand or the sole of the foot.

   _____

   _____

   _____

   _____

   _____

   _____

LAB PRACTICAL Integumentary System

1. Identify the structure.

   _____

2. Identify the layer of the epidermis.

   _____

3. Identify the layer of the dermis.

   _____

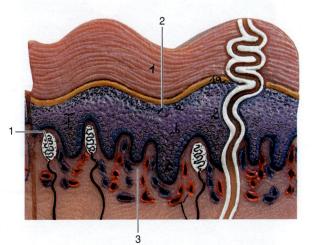

Copyright by SOMSO, 2010, www.somso.com

4. Identify the layer.

   _____

5. Identify the layer.

   _____

6. Identify the layer.

   _____

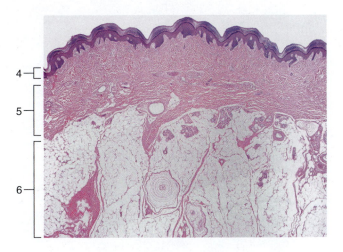

7. Identify the structure.

   _____

8. Identify the structures.

   _____

9. Identify the structure.

   _____

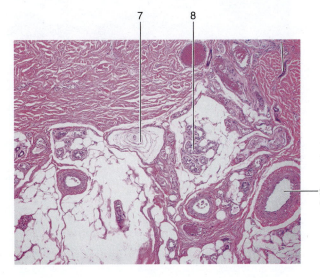

LAB PRACTICAL *continues*

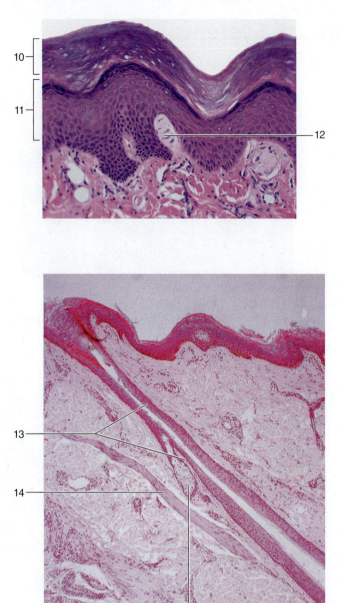

10. Identify the layer of the epidermis.

_____

11. Identify the layer of the epidermis.

_____

12. Identify the structure.

_____

13. Identify the structure.

_____

14. Identify the structure.

_____

15. Identify the structure.

_____

16. Identify the structure.

_____

17. Identify the structure.

_____

18. Identify the structure.

_____

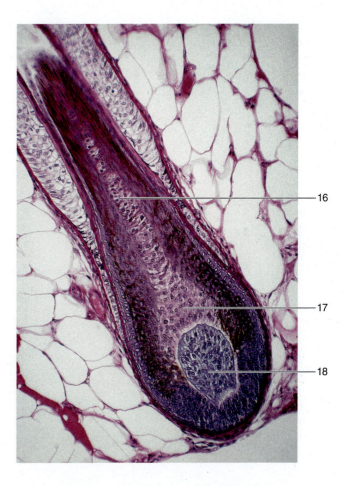

19. Identify the structure.

_____

20. Identify the structure.

_____

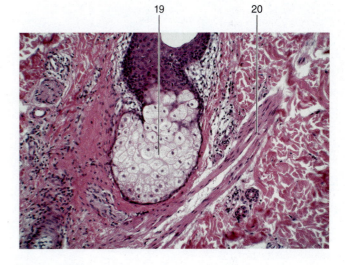

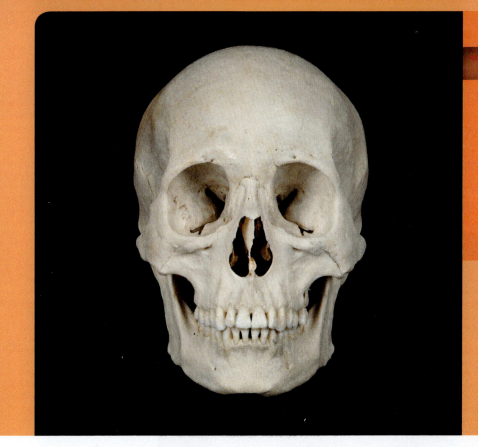

The Skeletal System

STUDENT OBJECTIVES

AXIAL SKELETON

1. Identify the major parts of the axial skeleton.
2. Identify the bones and important markings of the skull.
3. Identify the bones that make up the orbit, nasal cavity, and paranasal sinuses.
4. Describe the general structure of the vertebral column. List the number of cervical, thoracic, and lumbar vertebrae.
5. Identify the atlas and the axis.
6. Identify the distinguishing characteristics of cervical, thoracic, and lumbar vertebrae.
7. Identify the important markings and openings of the sacrum.
8. Identify the bones and important markings of the sternum.
9. Identify the important markings of a rib.

APPENDICULAR SKELETON

10. Identify the major parts of the appendicular skeleton.
11. Identify the bones and important markings of the pectoral girdle.
12. Identify the bones and important markings of the upper limb.
13. Identify the bones and important markings of the pelvic girdle.
14. Describe the differences between a male and a female pelvis.
15. Identify the bones and important markings of the lower limb.

JOINTS

16. Identify the bones, ligaments, tendons, and other associated structures of the following joints: intervertebral disc, shoulder, knee, elbow, articulated pelvis, and hip.

TISSUES OF THE SKELETAL SYSTEM

17. Identify the following cartilage tissues and their parts: hyaline cartilage, fibrocartilage, and elastic cartilage.
18. Identify the following bone tissues and their parts: compact bone and spongy bone.
19. Identify the epiphyseal plate and its zones.

AXIAL SKELETON

SELF REVIEW Skull

Exercise 2.1 **Skull, Anterior View**

GO TO〉 HUMAN CADAVER > AXIAL SKELETON > SKULL > SELF REVIEW > IMAGES 1 AND 2

- *Mouse over the images to locate and label the structures indicated below. Click on the structures to hear their pronunciations.*

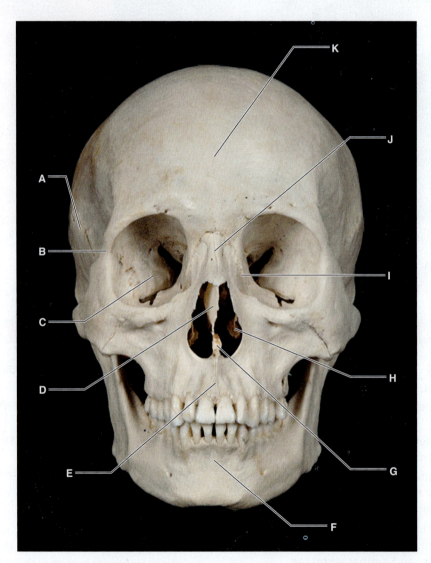

A. (bone) _____

B. (bone) _____

C. (bone) _____

D. (bone) _____

E. (bone) _____

F. (bone) _____

G. (bone) _____

H. _____

I. (bone) _____

J. (bone) _____

K. (bone) _____

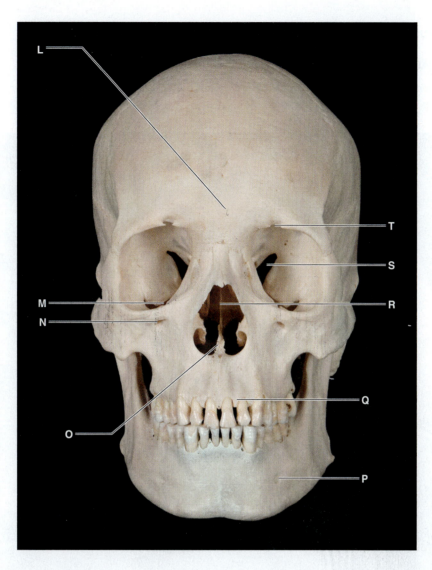

L. <u>Frontal Bone</u>

M. <u>Lacrimal Bone</u>

N. <u>Infraorbital Foramen</u>

O. <u>Anterior Nasal Spine</u>

P. <u>Mental Foramen</u>

Q. <u>Alveolar Process</u>

R. <u>Vomer / Orbital Fissure</u>

S. <u>Superior Foramen</u>

T. <u>Supraorbital Margin</u>

Exercise 2.2 **Skull, Lateral View**

GO TO HUMAN CADAVER > AXIAL SKELETON > SKULL > SELF REVIEW > IMAGE 4

• *Mouse over the image to locate and label the structures indicated below. Click on the structures to hear their pronunciations.*

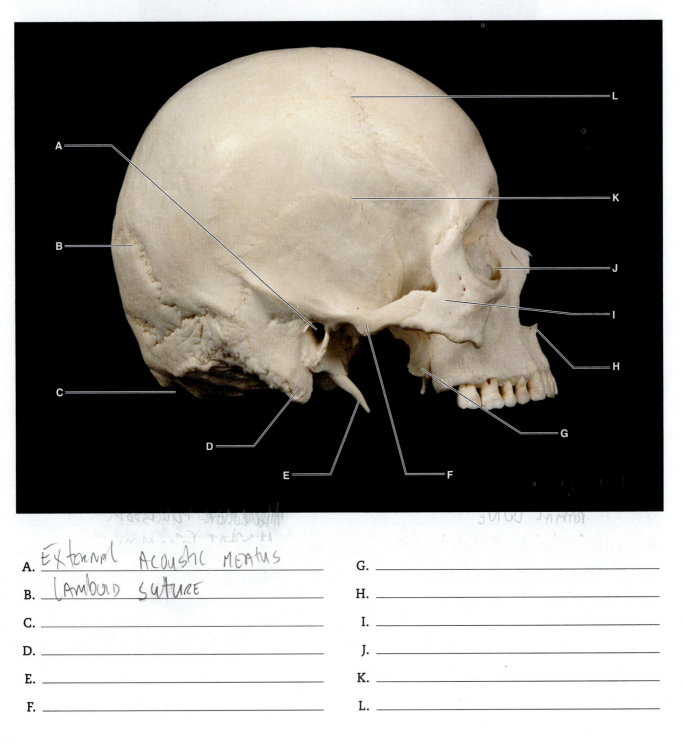

A. Exturnal Acoustic Meatus

B. Lambuid Suture

C. _____

D. _____

E. _____

F. _____

G. _____

H. _____

I. _____

J. _____

K. _____

L. _____

Exercise 2.3   Skull, Posterior View

GO TO > HUMAN CADAVER > AXIAL SKELETON > SKULL > SELF REVIEW > IMAGE 5

- *Mouse over the image to locate and label the structures indicated below. Click on the structures to hear their pronunciations.*

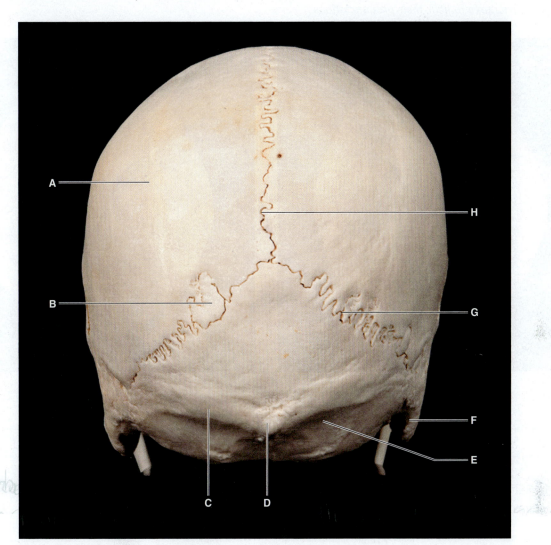

A. (bone) _____	E. (bone) _____
B. (bone) _____	F. (bone) _____
C. _____	G. _____
D. _____	H. _____

Exercise 2.4   **Skull, Posterosuperior View of the Bones and Structures of the Cranial Cavity**

GO TO ⟩ HUMAN CADAVER > AXIAL SKELETON > SKULL > SELF REVIEW > IMAGES 7 AND 8

- *Mouse over the images to locate and label the structures indicated below. Click on the structures to hear their pronunciations.*

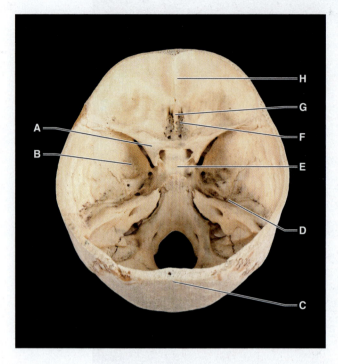

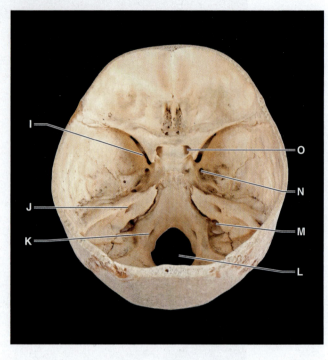

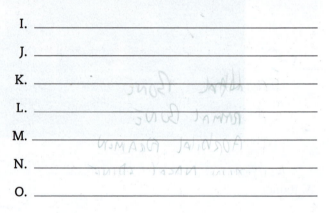

A. _____

B. _____

C. (bone) _____

D. (bone) _____

E. _____

F. _____

G. _____

H. (bone) _____

I. _____

J. _____

K. _____

L. _____

M. _____

N. _____

O. _____

Exercise 2.5 Skull, Superior View of Cranial Cavity Openings

GO TO ▷ HUMAN CADAVER > AXIAL SKELETON > SKULL > SELF REVIEW > IMAGE 10

• *Mouse over the image to locate and label the openings indicated below. Click on the structures to hear their pronunciations.*

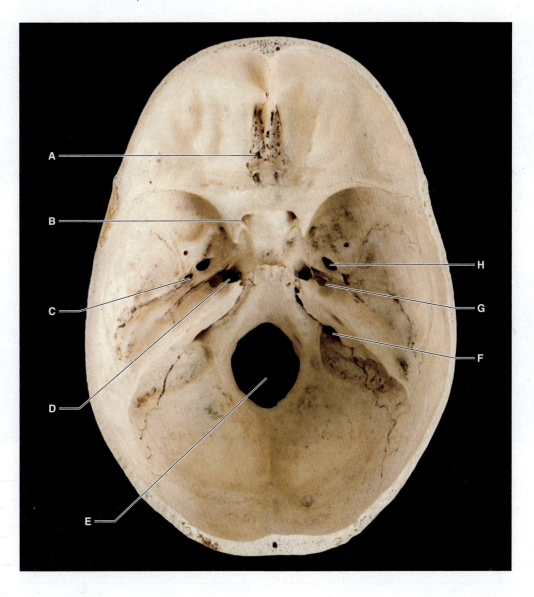

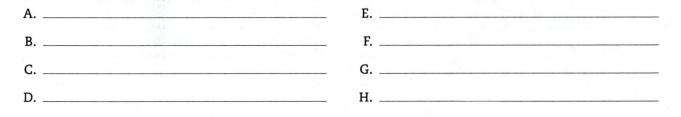

A. _____   E. _____

B. _____   F. _____

C. _____   G. _____

D. _____   H. _____

Exercise 2.6 **Skull, Inferior View**

GO TO ⟩ HUMAN CADAVER > AXIAL SKELETON > SKULL > SELF REVIEW > IMAGES 11 AND 12

- *Mouse over the images to locate and label the structures indicated below. Click on the structures to hear their pronunciations.*

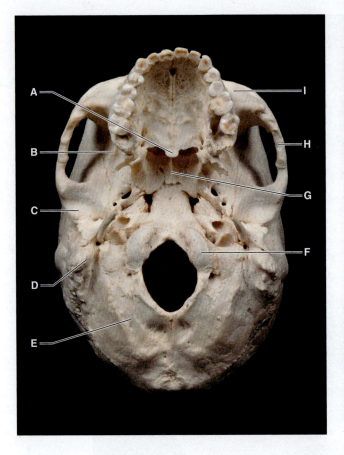 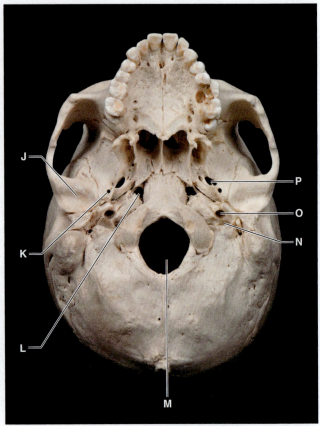

A. (bone) _____

B. (bone) _____

C. (bone) _____

D. _____

E. (bone) _____

F. _____

G. (bone) _____

H. (bone) _____

I. (bone) _____

J. _____

K. _____

L. _____

M. _____

N. _____

O. _____

P. _____

Exercise 2.7 Bones of the Orbit

GO TO > HUMAN CADAVER > AXIAL SKELETON > SKULL > SELF REVIEW > IMAGE 34

- *Mouse over the image to locate and label the structures indicated below. Click on the structures to hear their pronunciations.*

A. _____

B. _____

C. _____

D. _____

E. _____

F. _____

G. _____

Exercise 2.8 Hyoid

GO TO > HUMAN CADAVER > AXIAL SKELETON > SKULL > SELF REVIEW > IMAGE 35

- *Mouse over the image to locate and label the structures indicated below. Click on the structures to hear their pronunciations.*

A. _____

B. _____

C. _____

SELF REVIEW | Vertebral Column

Exercise 2.9 **Vertebral Column, Lateral View**

GO TO ⟩ HUMAN CADAVER > AXIAL SKELETON > VERTEBRAL COLUMN > SELF REVIEW > IMAGE 1

- *Mouse over the image to locate and label the structures and regions of the vertebral column indicated below. Click on the structures to hear their pronunciations.*

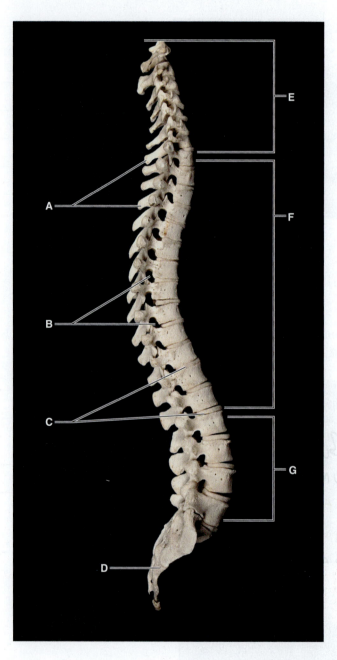

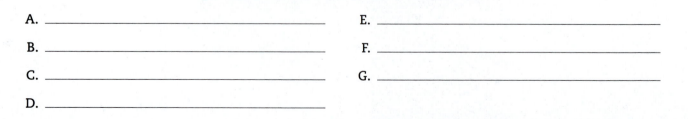

A. _____

B. _____

C. _____

D. _____

E. _____

F. _____

G. _____

Exercise 2.10   Atlas, Superior View

GO TO ▷ HUMAN CADAVER > AXIAL SKELETON > VERTEBRAL COLUMN > SELF REVIEW > IMAGE 3

- *Mouse over the image to locate and label the structures indicated below. Click on the structures to hear their pronunciations.*

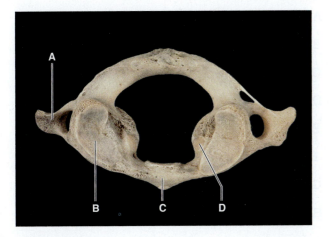

A. _____

B. _____

C. _____

D. _____

Exercise 2.11   Axis, Superior and Inferior Views

GO TO ▷ HUMAN CADAVER > AXIAL SKELETON > VERTEBRAL COLUMN > SELF REVIEW > IMAGES 5 AND 6

- *Mouse over the images to locate and label the structures indicated below. Click on the structures to hear their pronunciations.*

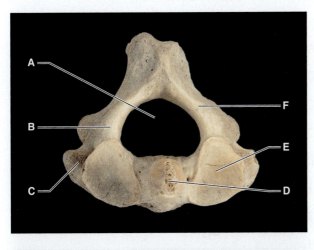

A. _____

B. _____

C. _____

D. _____

E. _____

F. _____

G. _____

H. _____

I. _____

J. _____

Exercise 2.12   **Cervical Vertebra, Superior View**

GO TO⟩ HUMAN CADAVER > AXIAL SKELETON > VERTEBRAL COLUMN > SELF REVIEW > IMAGE 7

- *Mouse over the image to locate and label the structures indicated below. Click on the structures to hear their pronunciations.*

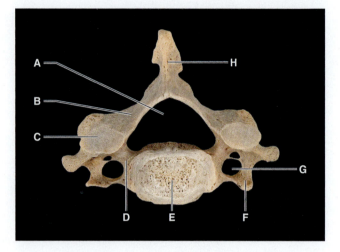

A. _____

B. _____

C. _____

D. _____

E. _____

F. _____

G. _____

H. _____

Exercise 2.13   **Thoracic Vertebra, Lateral View, Right Side**

GO TO⟩ HUMAN CADAVER > AXIAL SKELETON > VERTEBRAL COLUMN > SELF REVIEW > IMAGE 11

- *Mouse over the image to locate and label the structures indicated below. Click on the structures to hear their pronunciations.*

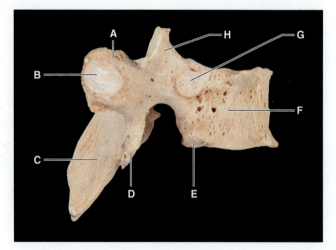

A. _____

B. _____

C. _____

D. _____

E. _____

F. _____

G. _____

H. _____

Exercise 2.14 **Sacrum and Coccyx, Posterior View**

GO TO HUMAN CADAVER > AXIAL SKELETON > VERTEBRAL COLUMN > SELF REVIEW > IMAGE 24

> • *Mouse over the image to locate and label the structures indicated below. Click on the structures to hear their pronunciations.*

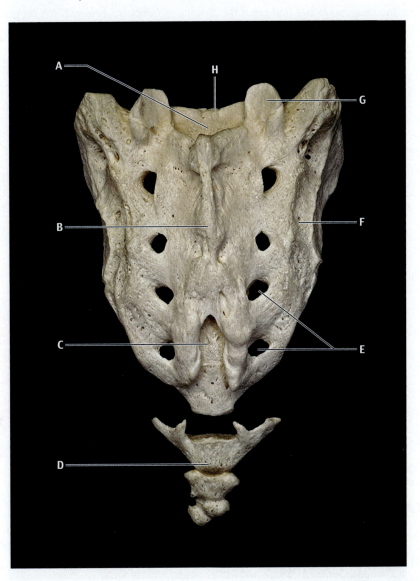

A. _____   E. _____

B. _____   F. _____

C. _____   G. _____

D. _____   H. _____

SELF REVIEW | Thoracic Cage

Exercise 2.15   **Sternum, Anterior View**

GO TO ⟩ HUMAN CADAVER > AXIAL SKELETON > THORACIC CAGE > SELF REVIEW > IMAGE 3

- *Mouse over the image to locate and label the structures indicated below. Click on the structures to hear their pronunciations.*

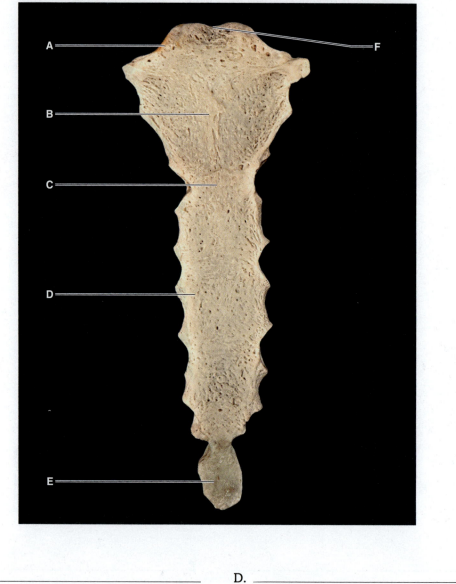

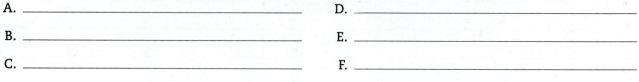

A. _____   D. _____

B. _____   E. _____

C. _____   F. _____

Exercise 2.16 **Rib, Inferior View, Right Side**

GO TO HUMAN CADAVER > AXIAL SKELETON > THORACIC CAGE > SELF REVIEW > IMAGE 5

- *Mouse over the image to locate and label the structures indicated below. Click on the structures to hear their pronunciations.*

A. _____   D. _____

B. _____   E. _____

C. _____   F. _____

Bone Rotation Questions–Axial Skeleton

GO TO ▷ HUMAN CADAVER > AXIAL SKELETON > SKULL > SELF REVIEW > IMAGE 1

- *Click Rotate to rotate the skull and determine the views of the images below.*

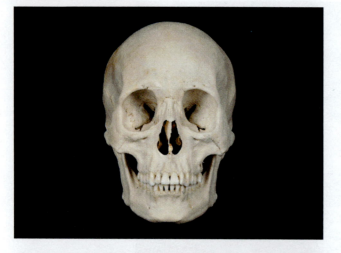

A. _____

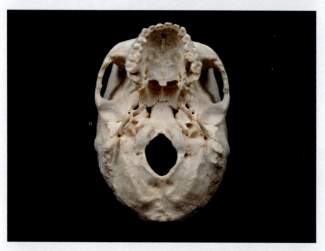

B. _____

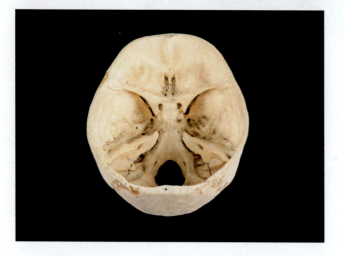

C. _____

QUIZ) Axial Skeleton

Answer the following questions using information from PAL 3.1 as well as other course materials including your textbook, lecture, and lab notes.

I. Check Your Understanding

1. The zygomatic process is part of which bone?

_____

2. Which bones are joined by the sagittal suture?

_____

3. The styloid process is found in which region of the temporal bone?

_____

4. The mandibular condyle articulates with which structure and bone?

_____

5. Name five of the bones found comprising the orbit of the skull.

_____

_____

6. Which two bones make up the zygomatic arch?

_____

7. Where is the mastoid process palpated externally?

_____

8. Which bones form the greater part of the nasal septum?

_____

9. Check yes or no to indicate whether the following openings are part of the temporal bone.

a. external acoustic meatus   yes ❑   no ❑

b. jugular foramen   yes ❑   no ❑

c. internal acoustic meatus   yes ❑   no ❑

d. carotid canal   yes ❑   no ❑

e. foramen spinosum   yes ❑   no ❑

f. foramen ovale   yes ❑   no ❑

g. foramen rotundum   yes ❑   no ❑

h. optic canal   yes ❑   no ❑

10. Indicate the correct number of vertebrae found in each region of the vertebral column:

cervical _____

thoracic _____

lumbar _____

sacrum _____

coccyx _____

11. Which features are unique to all cervical vertebrae?

_____

12. Which features are unique to thoracic vertebrae?

_____

13. Is the body of a vertebra located in the anterior or posterior portion of the column?

_____

14. How many bones make up the sternum? Name them.

_____

_____

15. How many pairs of ribs are there?

_____

16. What is the difference between true ribs and false ribs?

_____

_____

_____

_____

BEYOND PAL

17. Which bone of the body has no point of contact with another bone? Is this bone part of the axial or appendicular skeleton?

_____

_____

18. Which important nerves are transmitted through the cribriform plate and optic canals?

_____

_____

19. Which bones contain the paranasal sinuses? What functions do the sinuses serve?

Bones

_____

_____

Functions

_____

_____

20. The main sources of oxygenated blood supplying the brain are the internal carotid and vertebral arteries. Deoxygenated blood leaves the brain by way of the internal jugular veins. What openings in the axial skeleton transmit these vital blood vessels?

Vertebral arteries

_____

Internal carotid arteries

_____

Internal jugular veins

_____

II. Apply What You Learned

1. Briefly describe the anatomy of a normal intervertebral disc and a herniated one. What symptoms might a herniated disc produce?

Normal

_____

_____

Herniated

_____

_____

_____

2. The primary and secondary curvatures of the vertebral column are formed as a result of distinct developmental milestones. What are these milestones?

Primary

_____

_____

Secondary

_____

_____

APPENDICULAR SKELETON

SELF REVIEW ⟩ Pectoral Girdle

Exercise 2.17 **Clavicle, Superior View, Right Side**

GO TO ⟩ HUMAN CADAVER > APPENDICULAR SKELETON > PECTORAL GIRDLE > SELF REVIEW > IMAGE 5

- *Mouse over the image to locate and label the structures indicated below. Click on the structures to hear their pronunciations.*

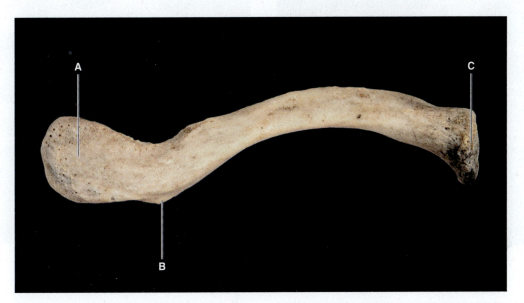

A. _____   C. _____

B. _____

Exercise 2.18 **Scapula, Anterior and Posterior Views, Right Side**

GO TO HUMAN CADAVER > APPENDICULAR SKELETON > PECTORAL GIRDLE > SELF REVIEW > IMAGES 8 AND 10

> • *Mouse over the images to locate and label the structures indicated below. Click on the structures to hear their pronunciations.*

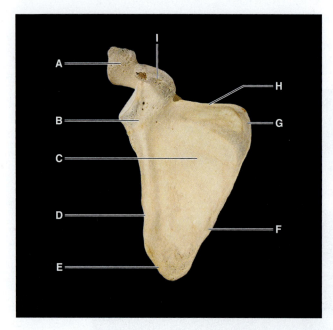
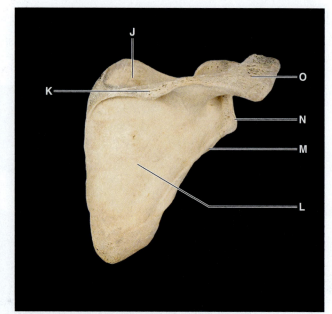

A. _____ I. _____

B. _____ J. _____

C. _____ K. _____

D. _____ L. _____

E. _____ M. _____

F. _____ N. _____

G. _____ O. _____

H. _____

SELF REVIEW | Upper Limb

Exercise 2.19 **Humerus, Anterior and Posterior Views, Right Side**

GO TO HUMAN CADAVER > APPENDICULAR SKELETON > UPPER LIMB > SELF REVIEW > IMAGES 1 AND 5

- *Mouse over the images to locate and label the structures indicated below. Click on the structures to hear their pronunciations.*

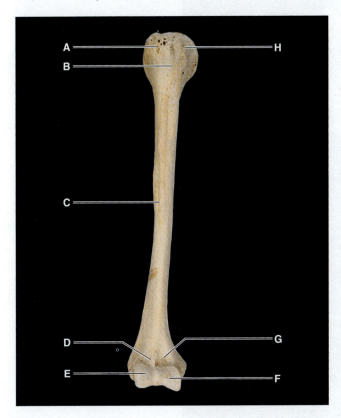
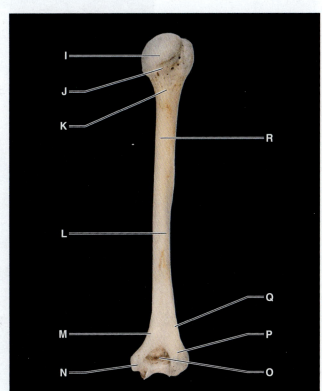

A. _____   J. _____

B. _____   K. _____

C. _____   L. _____

D. _____   M. _____

E. _____   N. _____

F. _____   O. _____

G. _____   P. _____

H. _____   Q. _____

I. _____   R. _____

Exercise 2.20   **Ulna, Lateral View, Right Side**

GO TO   HUMAN CADAVER > APPENDICULAR SKELETON > UPPER LIMB > SELF REVIEW > IMAGE 14

> • *Mouse over the image to locate and label the structures indicated below. Click on the structures to hear their pronunciations.*

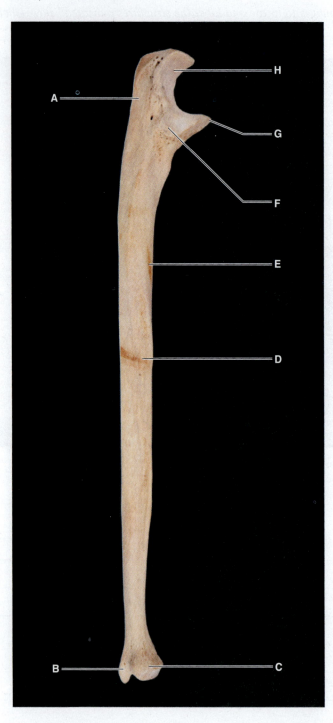

A. _____

B. _____

C. _____

D. _____

E. _____

F. _____

G. _____

H. _____

Exercise 2.21 Radius, Anterior View, Right Side

GO TO ⟩ HUMAN CADAVER > APPENDICULAR SKELETON > UPPER LIMB > SELF REVIEW > IMAGE 19

- *Mouse over the image to locate and label the structures indicated below. Click on the structures to hear their pronunciations.*

A. _____   D. _____

B. _____   E. _____

C. _____   F. _____

Exercise 2.22   Bones of the Hand, Anterior and Posterior Views, Right Side

GO TO > HUMAN CADAVER > APPENDICULAR SKELETON > UPPER LIMB > SELF REVIEW > IMAGES 26 AND 27

• *Mouse over the images to locate and label the bones indicated below. Click on the bones to hear their pronunciations.*

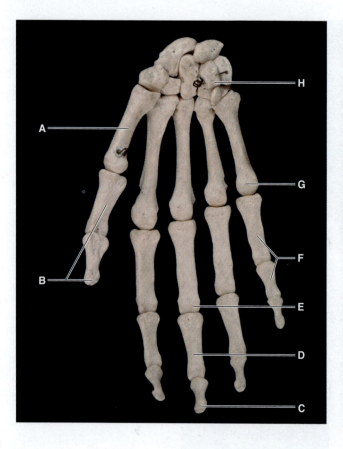

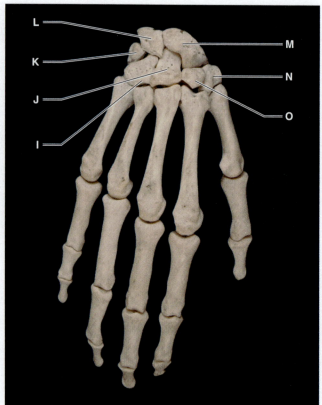

A. _____   I. _____

B. _____   J. _____

C. _____   K. _____

D. _____   L. _____

E. _____   M. _____

F. _____   N. _____

G. _____   O. _____

H. _____

SELF REVIEW | Pelvic Girdle

Exercise 2.23 **Articulated Male and Female Pelves, Anterior View**

GO TO ⟩ HUMAN CADAVER > APPENDICULAR SKELETON > PELVIC GIRDLE > SELF REVIEW > IMAGE 10

- *Mouse over the image to locate and label the structures indicated below. Click on the structures to hear their pronunciations.*

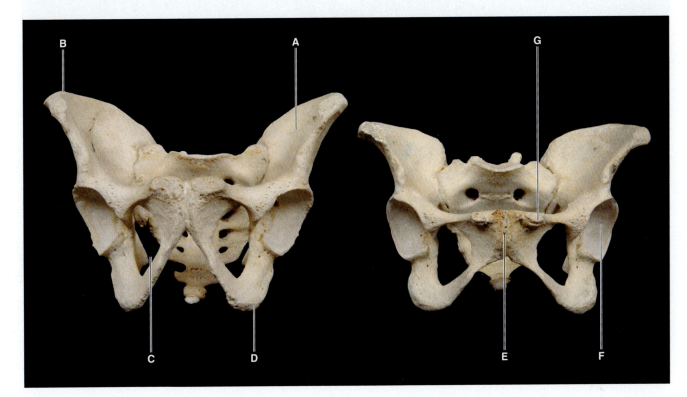

A. _____   E. _____

B. _____   F. _____

C. (opening) _____   G. _____

D. _____

Exercise 2.24   **Articulated Male and Female Pelves, Posterior View**

GO TO ⟩ HUMAN CADAVER > APPENDICULAR SKELETON > PELVIC GIRDLE > SELF REVIEW > IMAGE 12

- *Mouse over the image to locate and label the structures indicated below. Click on the structures to hear their pronunciations.*

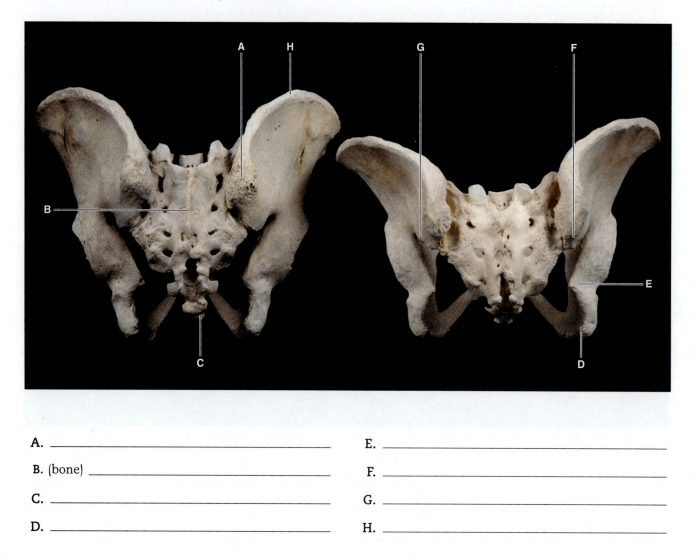

A. _____   E. _____

B. (bone) _____   F. _____

C. _____   G. _____

D. _____   H. _____

Exercise 2.25   ## Articulated Male and Female Pelves, Superior View

GO TO > HUMAN CADAVER > APPENDICULAR SKELETON > PELVIC GIRDLE > SELF REVIEW > IMAGE 13

• *Mouse over the image to locate and label the structures indicated below. Click on the structures to hear their pronunciations.*

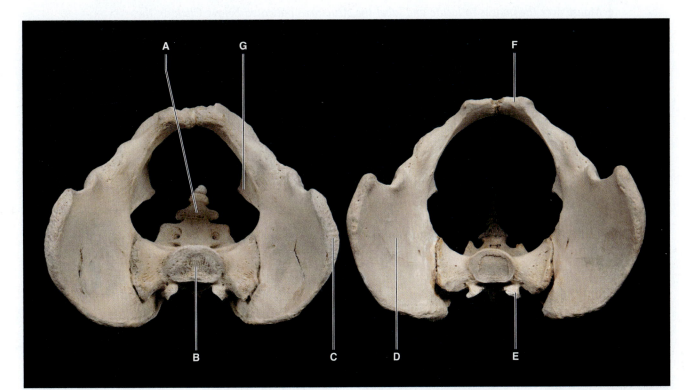

A. _____   E. _____

B. _____   F. _____

C. _____   G. _____

D. _____

SELF REVIEW | Lower Limb

Exercise 2.26 Femur, Anterior and Posterior Views, Right Side

GO TO ⟩ HUMAN CADAVER > APPENDICULAR SKELETON > LOWER LIMB > SELF REVIEW > IMAGES 1 AND 5

- *Mouse over the images to locate and label the structures indicated below. Click on the structures to hear their pronunciations.*

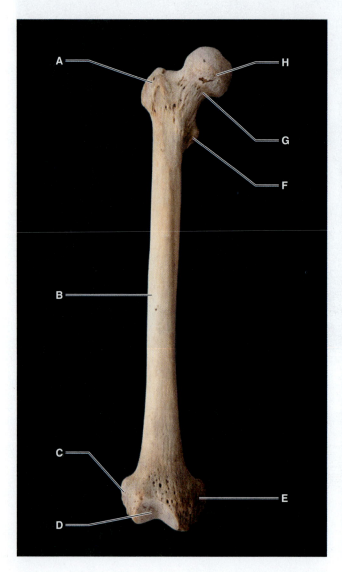

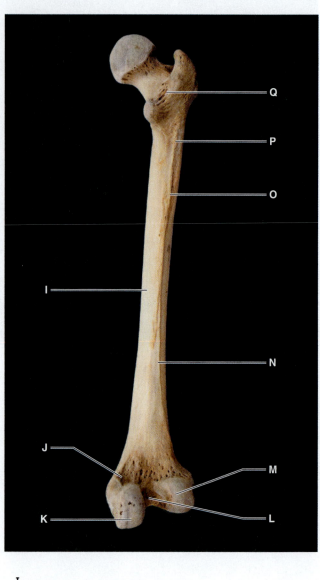

A. _____   I. _____

B. _____   J. _____

C. _____   K. _____

D. _____   L. _____

E. _____   M. _____

F. _____   N. _____

G. _____   O. _____

H. _____   P. _____

Q. _____

Exercise 2.27 Tibia, Anterior and Lateral Views, Right Side

GO TO ⟩ HUMAN CADAVER > APPENDICULAR SKELETON > LOWER LIMB > SELF REVIEW > IMAGES 19 AND 21

- *Mouse over the images to locate and label the structures indicated below. Click on the structures to hear their pronunciations.*

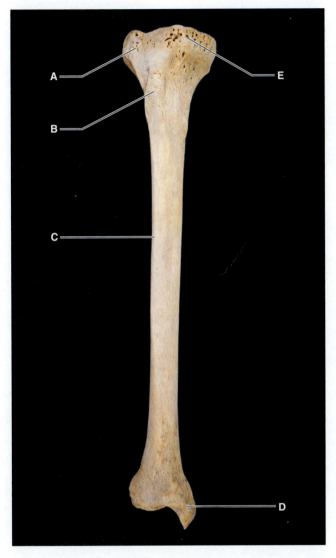

A. _____   E. _____

B. _____   F. _____

C. _____   G. _____

D. _____   H. _____

Exercise 2.28   Fibula, Anterior View, Right Side

GO TO > HUMAN CADAVER > APPENDICULAR SKELETON > LOWER LIMB > SELF REVIEW > IMAGE 25

- *Mouse over the image to locate and label the bones indicated below. Click on the structures to hear their pronunciations.*

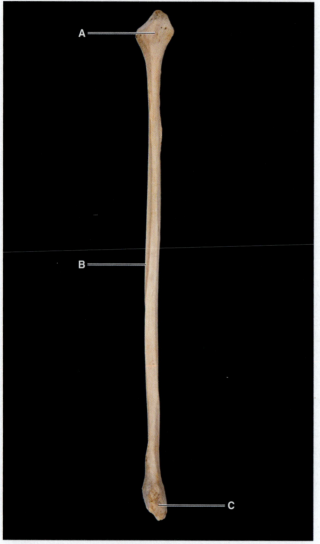

A. _____

B. _____

C. _____

Exercise 2.29 **Bones of the Foot, Superior View, Right Side**

GO TO HUMAN CADAVER > APPENDICULAR SKELETON > LOWER LIMB > SELF REVIEW > IMAGE 33

* *Mouse over the image to locate and label the bones indicated below. Click on the bones to hear their pronunciations.*

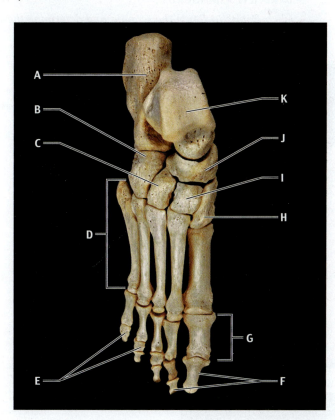

A. _____

B. _____

C. _____

D. _____

E. _____

F. _____

G. _____

H. _____

I. _____

J. _____

K. _____

Bone Rotation Questions–Appendicular Skeleton

Question 1

GO TO HUMAN CADAVER > APPENDICULAR SKELETON > PECTORAL GIRDLE > SELF REVIEW > IMAGE 10

- *Click Rotate to rotate the scapula and determine the views of the images below.*

A. _____

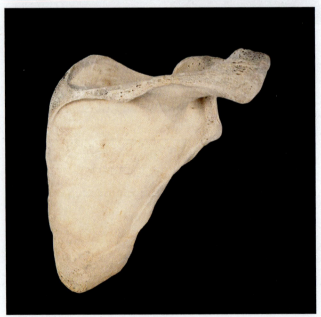

B. _____

Question 2

GO TO HUMAN CADAVER > APPENDICULAR SKELETON > LOWER LIMB > SELF REVIEW >
IMAGE 1

- *Click Rotate to rotate the femur and determine the views of the images below.*

A. _____   B. _____   C. _____

Question 3

GO TO HUMAN CADAVER > APPENDICULAR SKELETON > LOWER LIMB > SELF REVIEW >
IMAGE 19

- *Click Rotate to rotate the tibia and determine the views of the images below.*

A. _____ B. _____ C. _____

QUIZ | Appendicular Skeleton

Answer the following questions using information from PAL 3.1 as well as other course materials including your textbook, lecture, and lab notes.

I. Check Your Understanding

1. Which structure of the scapula articulates with the clavicle?

_____

2. Which bone and structure articulates with the glenoid fossa?

_____

3. Which bones articulate with the distal radius?

_____

_____

4. What are the eight carpal bones?

_____

_____

_____

_____

5. When you lean on a table with your elbow, which structure of the ulna do you press on?

_____

6. In correct anatomical position, which bone of the forearm is lateral?

_____

7. Which bone articulates with the head of the 5th proximal phalanx?

   a. 5th middle phalanx

   b. 5th distal phalanx

   c. 5th metacarpal

   d. hamate

   e. capitate

8. Match the structures below with the bones on which they are found. (Choose from the following bones: **clavicle**, **scapula**, **humerus**, **ulna**, and/or **radius**.)

   a. greater tubercle   _____

   b. conoid tubercle   _____

   c. radial groove   _____

   d. coracoid process   _____

   e. capitulum   _____

   f. trochlear notch   _____

   g. styloid process   _____

   h. infraspinous fossa _____

9. Which three bones fuse to make a hip (coxal) bone?

_____

_____

10. Which bone and structure articulate with the acetabulum?

_____

11. Which bone articulates with the proximal fibula?

_____

12. Which bone makes up the heel of your foot?

_____

13. Which structure of the ischium do you "sit" on?

_____

14. Where does the patella articulate with the femur?

_____

15. Which bone articulates with the medial side of the cuboid bone?

   a. metatarsal

   b. medial cuneiform

   c. lateral cuneiform

   d. calcaneus

16. Match the structures below with the bones on which they are found. (Choose from the following bones: **femur**, **tibia**, **patella**, and/or **fibula**.)

    a. gluteal tuberosity _____

    b. intercondylar
       eminence   _____

    c. lateral malleolus   _____

    d. medial malleolus   _____

    e. apex   _____

    f. adductor tubercle _____

    g. fibular notch   _____

17. Identify three distinctive features of the female pelvis.

    _____

    _____

    _____

II. Apply What You Learned

1. Describe the anatomy of the carpal tunnel. What is the clinical significance of this passageway?

    _____

    _____

    _____

    _____

    _____

    _____

    _____

2. Why would the deltoid tuberosity of a male body builder be greater than that of an age-matched, weight-matched male couch potato?

    _____

    _____

    _____

    _____

    _____

JOINTS

SELF REVIEW | Human Cadaver

Exercise 2.30 **Hip Joint**

GO TO > HUMAN CADAVER > JOINTS > SELF REVIEW > IMAGE 16

- *Mouse over the image to locate and label the structures indicated below. Click on the structures to hear their pronunciations.*

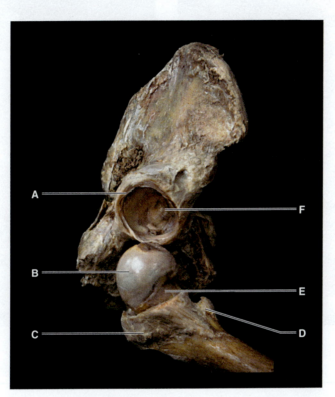

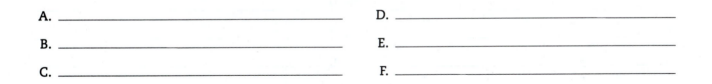

A. _____   D. _____

B. _____   E. _____

C. _____   F. _____

Exercise 2.31 **Knee Joint, Anterior Views**

GO TO > HUMAN CADAVER > JOINTS > SELF REVIEW > IMAGES 18 AND 20

- *Mouse over the images to locate and label the structures indicated below. Click on the structures to hear their pronunciations.*

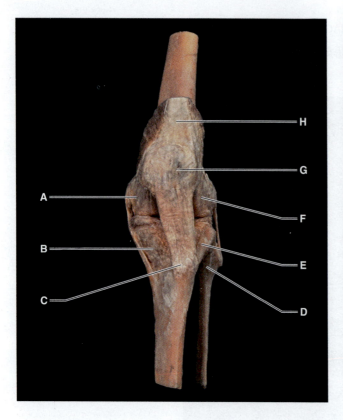
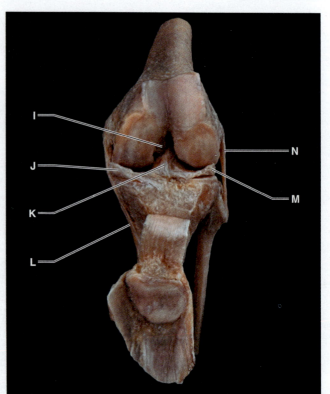

A. _____

B. _____

C. _____

D. _____

E. _____

F. _____

G. _____

H. _____

I. _____

J. _____

K. _____

L. _____

M. _____

N. _____

QUIZ] Human Cadaver

Answer the following questions using information from PAL 3.1 as well as other course materials including your textbook, lecture, and lab notes.

I. Check Your Understanding

1. Which bones articulate to form the knee joint?

_____

_____

2. For each structure listed, indicate whether it is part of the **hip** or **knee joint**.

a. medial meniscus   _____

b. posterior cruciate ligament _____

c. ligamentum teres   _____

d. iliofemoral ligament   _____

e. tibial collateral ligament _____

BEYOND PAL **3.** Functionally, the pubic symphysis is an amphiarthrosis. What structural type is it?

_____

4. For each joint listed, indicate whether it is a **synovial, cartilaginous,** or **fibrous joint**.

a. shoulder   _____

b. intervertebral disc   _____

c. carpometacarpal   _____

d. pubic symphysis   _____

e. hip   _____

f. sagittal suture   _____

g. knee   _____

h. temporomandibular   _____

i. epiphyseal plate (juvenile) _____

j. gomphosis   _____

II. Apply What You Learned

1. Maria is scheduled for a complete knee replacement. During this procedure, which bones are removed and replaced with synthetic materials?

_____

_____

_____

2. The patella, a bone of the knee joint, is the largest sesamoid bone in the body. What is a sesamoid bone? What is the functional role of the patella?

_____

_____

_____

_____

_____

_____

_____

_____

SELF REVIEW | Anatomical Models

Exercise 2.32 Shoulder Joint, Anterior and Posterior Views

GO TO > ANATOMICAL MODELS > JOINTS > SELF REVIEW > IMAGES 4 AND 5

- *Mouse over the images to locate and label the structures indicated below. Click on the structures to hear their pronunciations.*

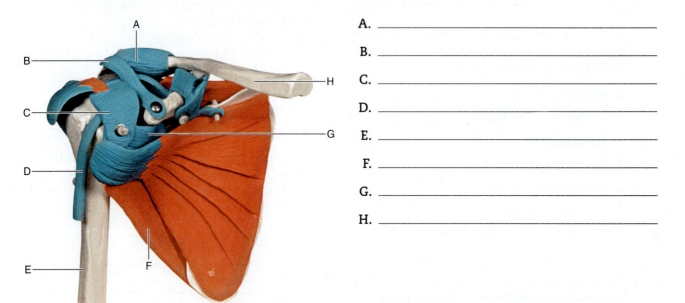

A. _____

B. _____

C. _____

D. _____

E. _____

F. _____

G. _____

H. _____

Model courtesy of Denoyer-Geppert, www.denoyer.com

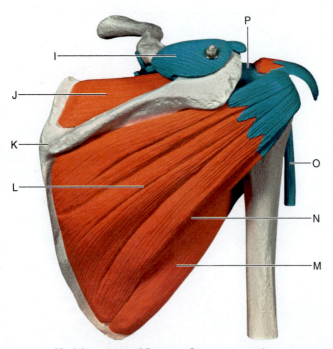

I. _____

J. _____

K. _____

L. _____

M. _____

N. _____

O. _____

P. _____

Model courtesy of Denoyer-Geppert, www.denoyer.com

Exercise 2.33   **Elbow Joint, Lateral and Medial Views**

GO TO ANATOMICAL MODELS > JOINTS > SELF REVIEW > IMAGES 7 AND 8

• *Mouse over the images to locate and label the structures indicated below. Click on the structures to hear their pronunciations.*

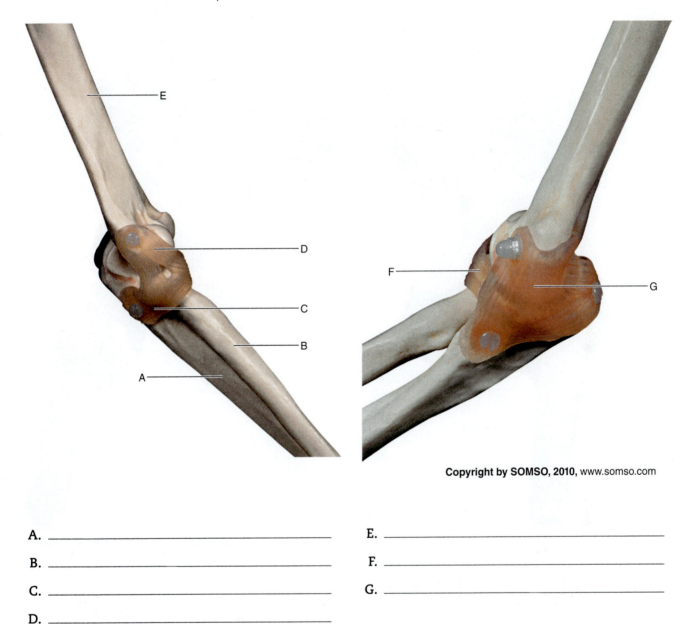

Copyright by **SOMSO, 2010,** www.somso.com

A. _____   E. _____

B. _____   F. _____

C. _____   G. _____

D. _____

Exercise 2.34 **Hip Joint, Anterior and Posterior Views**

GO TO ANATOMICAL MODELS > JOINTS > SELF REVIEW > IMAGES 10 AND 11

• *Mouse over the images to locate and label the structures indicated below. Click on the structures to hear their pronunciations.*

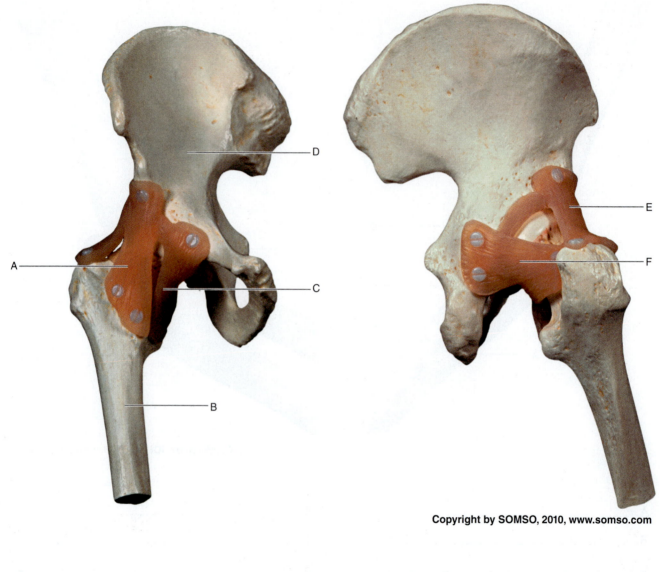

Copyright by SOMSO, 2010, www.somso.com

A. _____   D. _____

B. _____   E. _____

C. _____   F. _____

Exercise 2.35 Knee Joint, Anterior and Posterior Views

GO TO ANATOMICAL MODELS > JOINTS > SELF REVIEW > IMAGES 13 AND 14

- *Mouse over the images to locate and label the structures indicated below. Click on the structures to hear their pronunciations.*

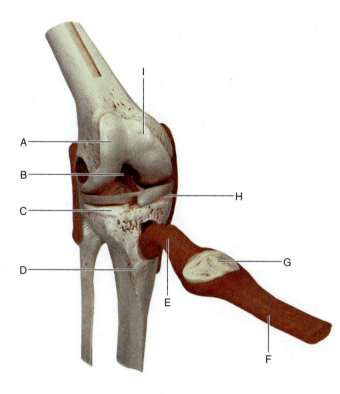

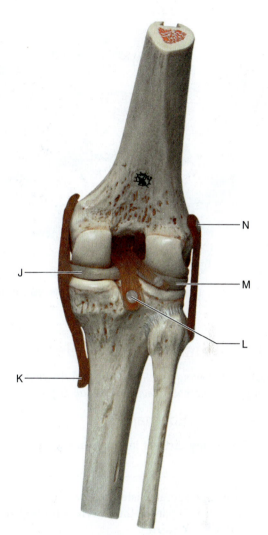

A. _____   H. _____

B. _____   I. _____

C. _____   J. _____

D. _____   K. _____

E. _____   L. _____

F. _____   M. _____

G. _____   N. _____

Exercise 2.36 **Female Pelvis**

GO TO ANATOMICAL MODELS > JOINTS > SELF REVIEW > IMAGE 17

• *Mouse over the image to locate and label the structures indicated below. Click on the structures to hear their pronunciations.*

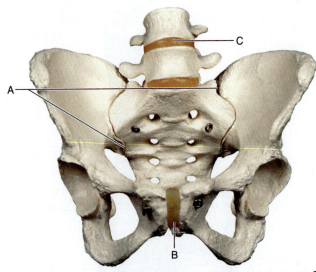

A. _____

B. _____

C. _____

LT-A61: Pelvic Skeleton, Female, 3B Scientific®

QUIZ | Anatomical Models

Answer the following questions using information from PAL 3.1 as well as other course materials including your textbook, lecture, and lab notes.

I. Check Your Understanding

1. How many distinct joints form the knee?

_____

2. For each structure listed below, indicate whether it is part of the **pelvic, elbow, hip,** and/or **knee joint**.

a. pubic symphysis _____

b. capitulum _____

c. lateral collateral ligament _____

d. acetabulum _____

e. anterior cruciate ligament _____

f. olecranon _____

g. tibial tuberosity _____

h. lateral meniscus _____

i. trochlear notch _____

j. acetabular labrum _____

k. medial condyle _____

BEYOND PAL **3.** The glenohumeral joint is a synovial joint. Which structural type is it?

_____

4. What three structures are typically damaged from a blow to the lateral aspect of the knee joint?

_____

_____

_____

5. For each synovial joint listed below, indicate its structural type (**plane, pivot, condyloid, saddle, hinge,** or **ball-and-socket**).

a. hip _____

b. patellofemoral _____

c. elbow _____

d. atlantooccipital _____

e. glenohumeral _____

f. radiocarpal _____

g. proximal radioulnar _____

h. intercarpal _____

i. ankle _____

j. atlantoaxial _____

k. carpometacarpal _____

II. Apply What You Learned

Gouty arthritis is a painful condition associated with synovial joints that occurs when crystals of uric acid (urate) precipitate out into the synovial fluid of the joint.

1. What structure, unique to synovial joints, is filled with synovial fluid?

_____

2. Individuals who are prone to gout are advised to avoid foods rich in purines, which are broken down into uric acid and transported through the blood. What structure associated with a synovial joint contains the blood vessels from which uric acid crystals precipitate?

_____

3. Why is the presence of uric acid crystals a painful problem?

_____

_____

_____

_____

TISSUES OF THE SKELETAL SYSTEM

SELF REVIEW | Histology

Exercise 2.37 **Hyaline Cartilage, Cross Section 400x**

GO TO HISTOLOGY >CONNECTIVE TISSUE > SELF REVIEW > IMAGE 19

- *Mouse over the image to locate and label the structures indicated below. Click on the structures to hear their pronunciations.*

A. _____

B. _____

C. _____

D. _____

Exercise 2.38 **Fibrocartilage, Pubic Symphysis, Cross Section 400x**

GO TO > HISTOLOGY >CONNECTIVE TISSUE > SELF REVIEW > IMAGE 22

- *Mouse over the image to locate and label the structures indicated below. Click on the structures to hear their pronunciations.*

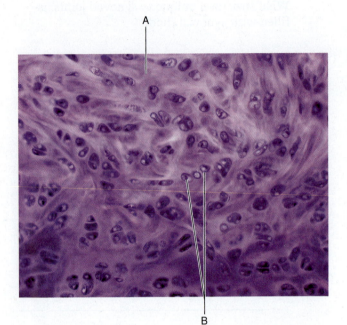

A. _____

B. _____

Exercise 2.39 **Elastic Cartilage, Epiglottis, Cross Section 400x**

GO TO > HISTOLOGY > CONNECTIVE TISSUE > SELF REVIEW > IMAGE 25

- *Mouse over the image to locate and label the structures indicated below. Click on the structures to hear their pronunciations.*

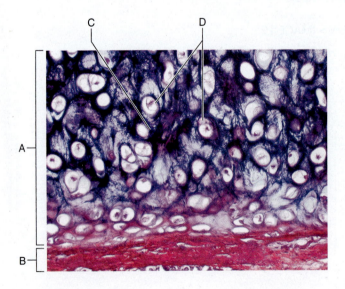

A. _____

B. _____

C. _____

D. _____

Exercise 2.40 Compact Bone, Ground, Cross Section 100x and 400x

GO TO HISTOLOGY > CONNECTIVE TISSUE > SELF REVIEW > IMAGES 27 and 29

- *Mouse over the images to locate and label the structures indicated below. Click on the structures to hear their pronunciations.*

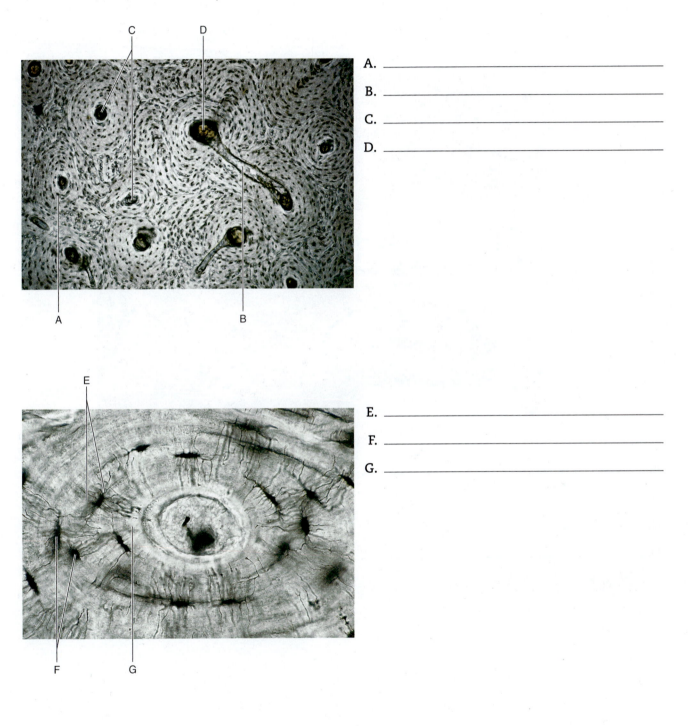

A. _____

B. _____

C. _____

D. _____

E. _____

F. _____

G. _____

Exercise 2.41 Developing Bone, Fetal Humerus, Cross Section 400x

GO TO HISTOLOGY > CONNECTIVE TISSUE > SELF REVIEW > IMAGE 32

- *Mouse over the image to locate and label the structures indicated below. Click on the structures to hear their pronunciations.*

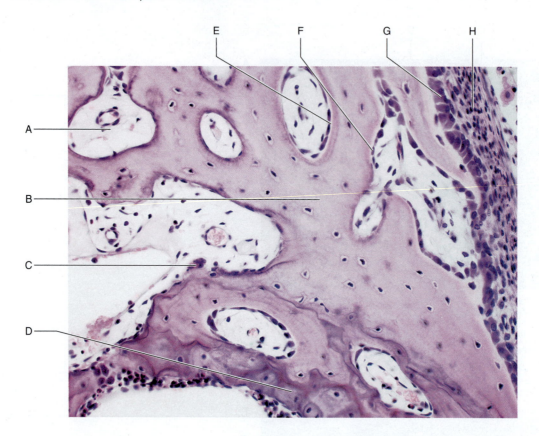

A. _____   E. _____

B. _____   F. (cells) _____

C. _____   G. (cells) _____

D. _____   H. _____

Exercise 2.42   **Epiphyseal Plate, Developing Bone, Longitudinal Section 15x and 200x**

GO TO HISTOLOGY > CONNECTIVE TISSUE > SELF REVIEW > IMAGES 36 AND 37

> • *Mouse over the images to locate and label the structures indicated below. Click on the structures to hear their pronunciations.*

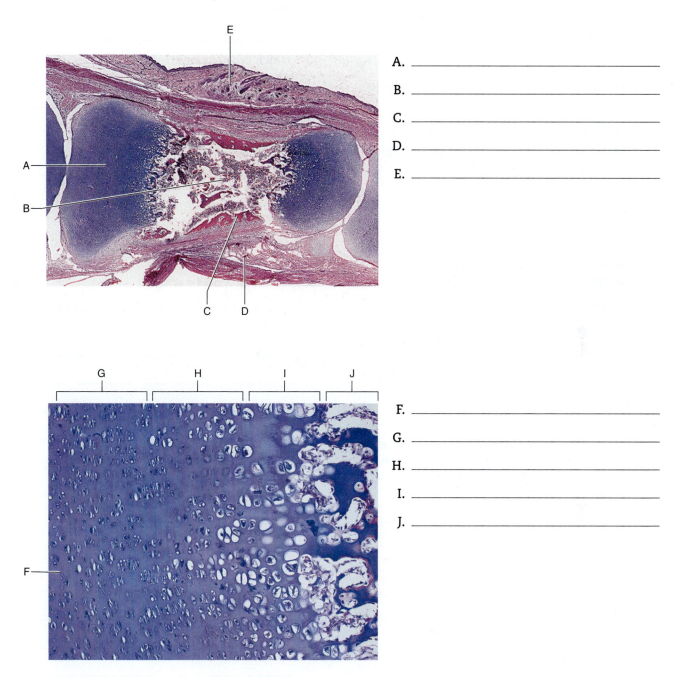

A. _____

B. _____

C. _____

D. _____

E. _____

F. _____

G. _____

H. _____

I. _____

J. _____

QUIZ] Histology

Answer the following questions using information from PAL 3.1 as well as other course materials including your textbook, lecture, and lab notes.

I. Check Your Understanding

1. What space within bone does an osteocyte occupy?

   _____

2. Which type of cartilage has a high percentage of elastic fibers?

   _____

3. Which region of the epiphyseal plate is closest to the ends of the long bone?

   a. growth zone

   b. resting zone

   c. hypertrophic zone

4. Which region of the epiphyseal plate is composed of older, enlarged chondrocytes?

   a. growth zone

   b. proliferation zone

   c. hypertrophic zone

5. What are the spicules of bone matrix called in spongy bone?

   _____

6. For each of the following structures, indicate if it is found in **compact bone only, spongy bone only,** or **BOTH spongy and compact bone.**

   a. lamella   _____

   b. central canal   _____

   c. osteocyte   _____

   d. lacuna   _____

   e. osteon   _____

   f. perforating canal   _____

   g. osteoclast   _____

BEYOND PAL 7. Which type of cartilage composes the tracheal rings?

   _____

8. Which type of cartilage is found in an intervertebral disc?

   _____

9. What is the function of osteoblasts and osteoclasts?

   _____

   _____

   _____

   _____

II. Apply What You Learned

1. Why is osteoporosis a particular concern for older women?

   _____

   _____

   _____

   _____

   _____

   _____

2. Damaged cartilage does not repair itself as well as damaged bone. Why?

   _____

   _____

   _____

   _____

   _____

   _____

LAB PRACTICAL Skeletal System

1. Identify the bone.

   _____

2. Identify the bone.

   _____

3. Identify the bone.

   _____

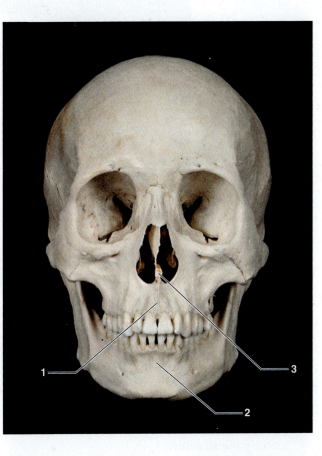

4. Identify the opening.

   _____

5. Identify the structure.

   _____

6. Identify the structure.

   _____

7. Identify the structure.

   _____

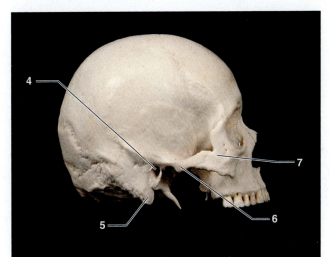

LAB PRACTICAL *continues*

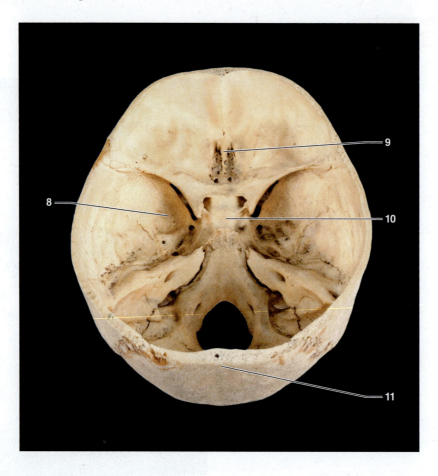

8. Identify the structure.

_____

9. Identify the structure.

_____

10. Identify the structure.

_____

11. Identify the bone.

_____

12. Identify the bone.

_____

13. Identify the opening.

_____

14. To which region does this vertebra belong?

_____

15. Identify the structure.

_____

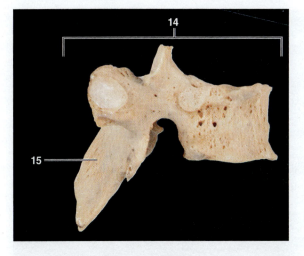

16. Identify the structure.

_____

17. Identify the structure.

_____

18. Identify the structure.

_____

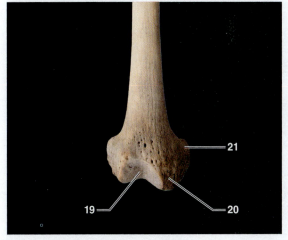

19. Identify the structure.

_____

20. Identify the structure.

_____

21. Identify the structure.

_____

LAB PRACTICAL _continues_

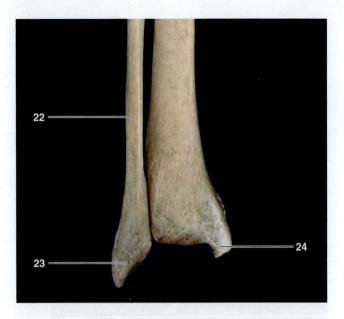

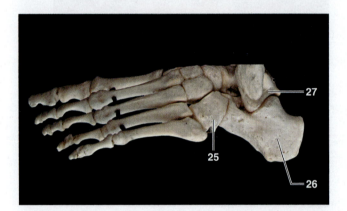

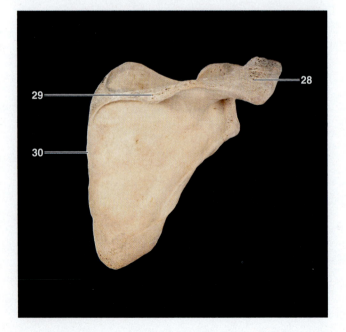

22. Identify the bone.

_____

23. Identify the structure.

_____

24. Identify the structure.

_____

25. Identify the bone.

_____

26. Identify the bone.

_____

27. Identify the bone.

_____

28. Identify the structure.

_____

29. Identify the structure.

_____

30. Identify the border.

_____

31. Identify the structure.

_____

32. Identify the structure.

_____

33. Identify the structure.

_____

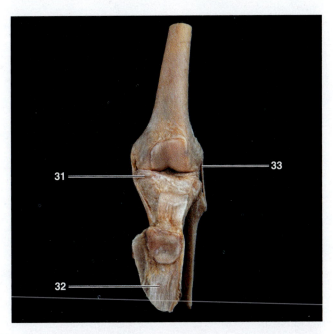

34. Identify the structure.

_____

35. Identify the structure.

_____

36. Identify the bone.

_____

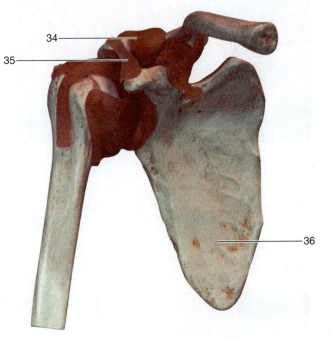

LAB PRACTICAL *continues*

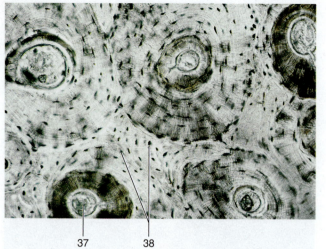

37

38

37. Identify the structure (space).

_____

38. Identify the structures (spaces).

_____

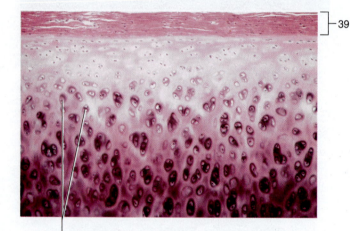

39

40

39. Identify the structure.

_____

40. Identify the cells.

_____

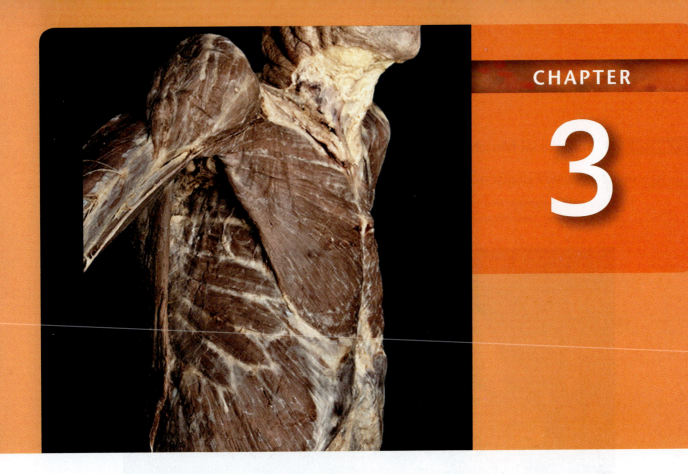

The Muscular System

MUSCLES OF THE HEAD AND NECK

SELF REVIEW | Human Cadaver

Exercise 3.1 **Superficial Muscles, Lateral View**

GO TO ⟩ HUMAN CADAVER > MUSCULAR SYSTEM > HEAD & NECK > SELF REVIEW > IMAGE 3

- *Mouse over the image to locate and label the muscles indicated below. Click on the muscles to hear their pronunciations.*

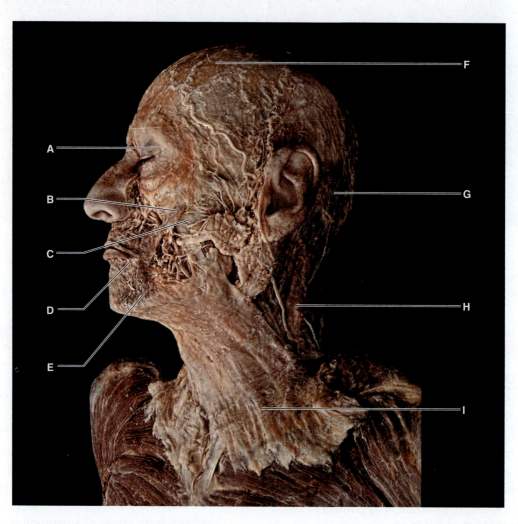

A. _____   F. _____

B. _____   G. _____

C. _____   H. _____

D. _____   I. _____

E. _____

Exercise 3.2 Deep Muscles, Lateral View

GO TO HUMAN CADAVER > MUSCULAR SYSTEM > HEAD & NECK > SELF REVIEW > IMAGES 4 AND 5

- *Mouse over the images and locate and label the muscles indicated below. Click on the muscles to hear their pronunciations.*

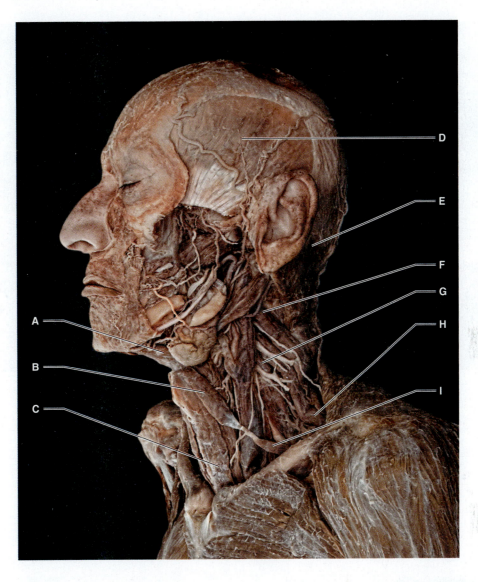

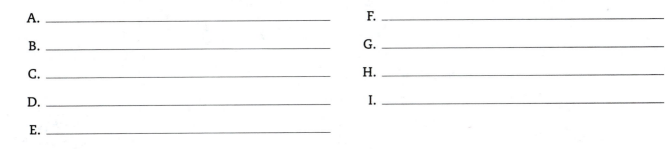

A. _____   F. _____

B. _____   G. _____

C. _____   H. _____

D. _____   I. _____

E. _____

Exercise 3.3 **Intermediate Muscles, Anterior View**

GO TO ➤ HUMAN CADAVER > MUSCULAR SYSTEM > HEAD & NECK > SELF REVIEW > IMAGE 23

- *Mouse over the image to locate and label the muscles indicated below. Click on the muscles to hear their pronunciations.*

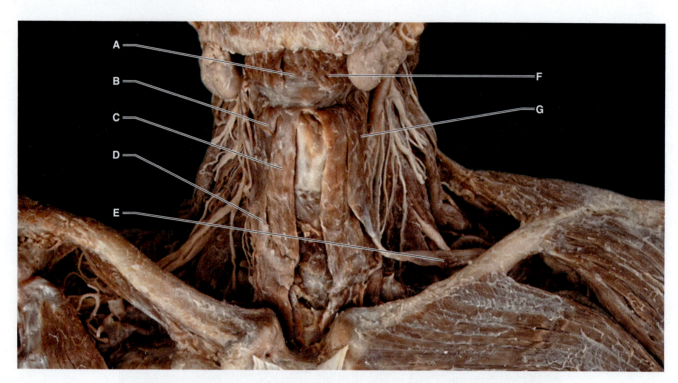

A. _____   E. _____

B. _____   F. _____

C. _____   G. _____

D. _____

Exercise 3.4   **Intermediate Muscles, Posterior View**

GO TO › HUMAN CADAVER > MUSCULAR SYSTEM > HEAD & NECK > SELF REVIEW > IMAGE 8

- *Mouse over the image to locate and label the muscles indicated below. Click on the muscles to hear their pronunciations.*

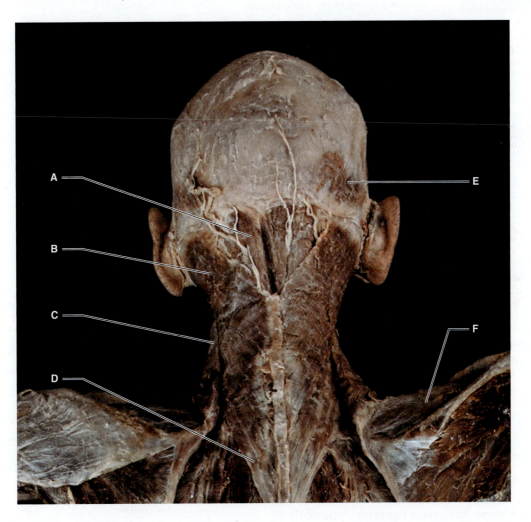

A. _____   D. _____

B. _____   E. _____

C. _____   F. _____

Animations Questions

GO TO HUMAN CADAVER > MUSCULAR SYSTEM > HEAD & NECK > SELF REVIEW

- *Go to the image number next to each muscle listed in the table below.*
- *Locate the muscle and double-click its label to view the animation.*
- *Complete the table below.*

MUSCLE	IMAGE NUMBER	ORIGIN	INSERTION	ACTION
Masseter	4			*closes jaw*
Temporalis	15	*temporal fossa*		

QUIZ Human Cadaver

Answer the following questions using information from PAL 3.1 as well as other course materials including your textbook, lecture, and lab notes.

I. Check Your Understanding

1. Name a muscle that has an origin on both the mastoid process of the temporal bone and the mandible.

   _____

2. What is the action of the omohyoid muscle?

   _____

   _____

3. Which nerve innervates the orbicularis oculi and orbicularis oris muscles?

   _____

4. Which muscle draws the corners of the mouth superiorly and laterally into a smile?

   _____

5. From which structure does the sternohyoid muscle originate?

   _____

6. Match the muscles, at left, with their correct insertion, at right.

   _____ temporalis   a. hyoid

   _____ mylohyoid   b. occipital

   _____ sternocleidomastoid   c. clavicle

   _____ trapezius   d. mandible

   _____ semispinalis capitis e. temporal

7. Match the muscles, at left, with their correct action, at right.

   _____ platysma   a. elevates larynx

   _____ thyrohyoid   b. depresses larynx

   _____ anterior scalene   c. depresses mandible

   _____ sternothyroid   d. elevates mandible

   _____ masseter   e. laterally flexes neck

II. Apply What You Learned

1. Bell's palsy is a condition that results in paralysis of facial muscles under the control of the facial nerve (cranial nerve VII). Identify four different muscles of facial expression whose function could be directly impacted in Bell's palsy.

_____

_____

_____

_____

2. Pain in the jaw is a disorder associated with the temporomandibular joint and is commonly referred to as TMJ. The muscles associated with moving the mandible, and consequently the temporomandibular joint, are the culprits. Name two major muscles of mastication that could contribute directly to pain caused by TMJ.

_____

_____

SELF REVIEW | Anatomical Models

Exercise 3.5   **Superficial Muscles, Anterolateral View**

GO TO ⟩ ANATOMICAL MODELS > MUSCULAR SYSTEM > HEAD & NECK > SELF REVIEW > IMAGE 1

- *Mouse over the image to locate and label the muscles indicated below. Click on the muscles to hear their pronunciations.*

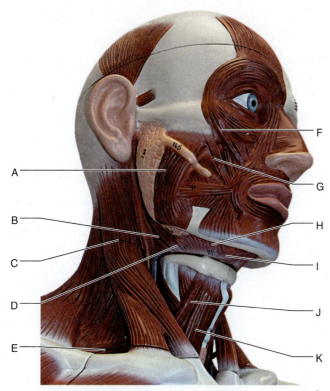

LT-VA16: Life-size muscle torso, 27-part, 3B Scientific®

A. _____

B. _____

C. _____

D. _____

E. _____

F. _____

G. _____

H. _____

I. _____

J. _____

K. _____

Exercise 3.6   Deep Muscles, Left Lateral View

GO TO ⟩ ANATOMICAL MODELS > MUSCULAR SYSTEM > HEAD & NECK > SELF REVIEW > IMAGE 6

• *Mouse over the image to locate and label the muscles indicated below. Click on the muscles to hear their pronunciations.*

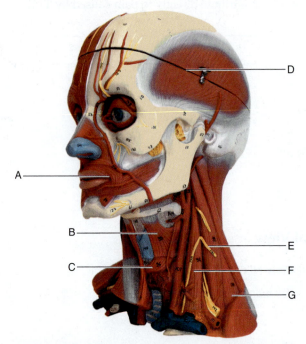

LT-CO5: Head and neck musculature, 5-part, 3B Scientic®

A. _____

B. _____

C. _____

D. _____

E. _____

F. _____

G. _____

Exercise 3.7   Superficial Muscles, Posterior View

GO TO ⟩ ANATOMICAL MODELS > MUSCULAR SYSTEM > HEAD & NECK > SELF REVIEW > IMAGE 8

• *Mouse over the image to locate and label the muscles indicated below. Click on the muscles to hear their pronunciations.*

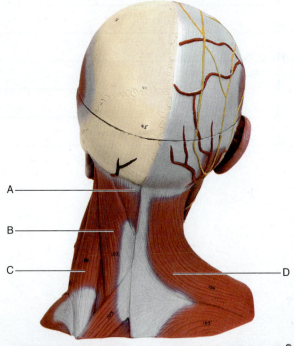

LT-CO5: Head dend neck musculature, 5-part, 3B Scientic®

A. _____

B. _____

C. _____

D. _____

Animations Questions

GO TO › ANATOMICAL MODELS > MUSCULAR SYSTEM > HEAD & NECK > SELF REVIEW

- *Go to the image number next to each muscle listed in the table below.*
- *Locate the muscle and double-click its label to view the animation.*
- *Complete the table below.*

MUSCLE	IMAGE NUMBER	ORIGIN	INSERTION	ACTION
Semispinalis capitis	8			
Splenius capitis	8			*bilateral contraction–extends neck* *unilateral contraction–extends, laterally flexes, and rotates head and neck to same side*

QUIZ〕Anatomical Models

Answer the following questions using information from PAL 3.1 as well as other course materials including your textbook, lecture, and lab notes.

I. Check Your Understanding

1. Which muscle depresses the hyoid and has an origin on the scapula?

_____

2. What is an action of the scalenes?

_____

_____

3. Which cranial nerve innervates the masseter muscle?

_____

4. Name a muscle that both depresses the hyoid and elevates the larynx.

_____

5. From which structure does the frontalis muscle originate?

_____

6. Match the muscles, at left, with their correct origin, at right.

_____ zygomaticus major

_____ digastric, posterior belly

_____ mylohyoid

_____ sternocleidomastoid

_____ orbicularis oculi

a. maxilla

b. temporal bone

c. zygomatic bone

d. mandible

e. manubrium

II. Apply What You Learned

Whiplash is a common injury associated with car accidents, but it can occur in any instance when the cervical region of the spine is subject to a force that causes hyperextension and/or hyperflexion. This "whipping" force can damage muscles, tendons, ligaments, nerves, and joints of the neck and nearby anatomical regions.

1. What major neck muscle, which you learned in this section, might be damaged when the neck is hyperextended?

_____

_____

2. What three major muscles associated with the neck, which you learned in this section, might be damaged when the neck is hyperflexed?

_____

_____

_____

MUSCLES OF THE TRUNK

SELF REVIEW | Human Cadaver

Exercise 3.8 Intermediate and Deep Muscles, Anterior View

GO TO ⟩ HUMAN CADAVER > MUSCULAR SYSTEM > TRUNK > SELF REVIEW > IMAGES 2 AND 21

- *Mouse over the images and locate and label the muscles indicated below. Click on the muscles to hear their pronunciations.*

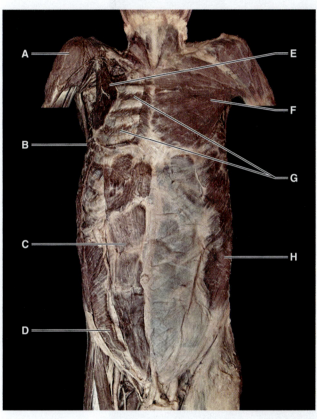

A. _____

B. _____

C. _____

D. _____

E. _____

F. _____

G. _____

H. _____

I. _____

J. _____

K. _____

Exercise 3.9 **Superficial Muscles, Lateral View**

GO TO HUMAN CADAVER > MUSCULAR SYSTEM > TRUNK > SELF REVIEW > IMAGE 4

• *Mouse over the image to locate and label the muscles indicated below. Click on the muscles to hear their pronunciations.*

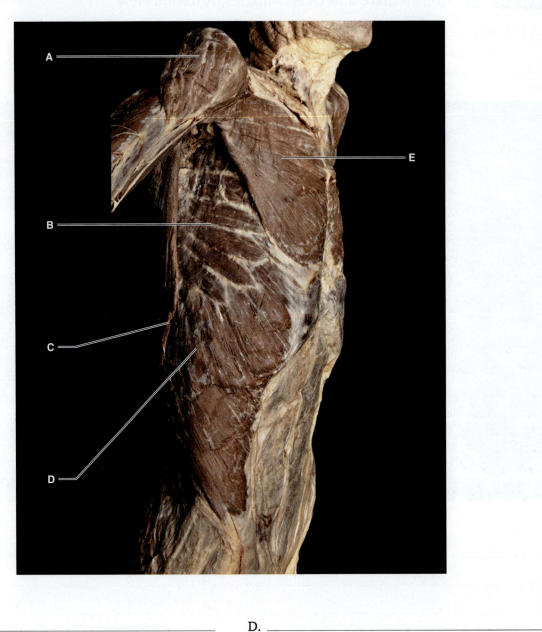

A. _____   D. _____

B. _____   E. _____

C. _____

Exercise 3.10   Posterior Muscles, Intermediate View

GO TO > HUMAN CADAVER > MUSCULAR SYSTEM > TRUNK > SELF REVIEW > IMAGE 7

• *Mouse over the image and locate and label the muscles indicated below. Click on the muscles to hear their pronunciations.*

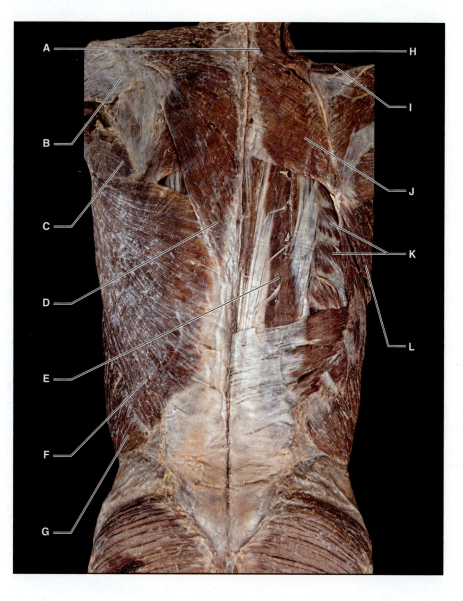

A. _____   G. _____

B. _____   H. _____

C. _____   I. _____

D. _____   J. _____

E. _____   K. _____

F. _____   L. _____

Exercise 3.11 **Posterior Muscles, Intermediate View**

GO TO ⟩ HUMAN CADAVER > MUSCULAR SYSTEM > TRUNK > SELF REVIEW > IMAGE 9

- *Mouse over the image and locate and label the muscles indicated below. Click on the muscles to hear their pronunciations.*

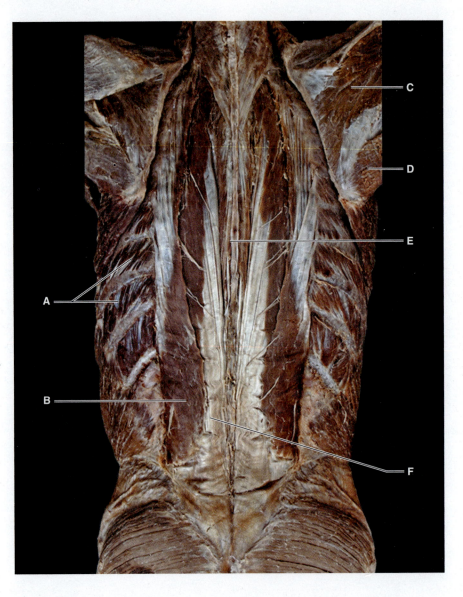

A. _____   D. _____

B. _____   E. _____

C. _____   F. _____

Animations Questions

GO TO ⟩ HUMAN CADAVER > MUSCULAR SYSTEM > TRUNK > SELF REVIEW
- *Go to the image number next to each muscle listed in the table below.*
- *Locate the muscle and double-click its label to view the animation.*
- *Complete the table below.*

MUSCLE	IMAGE NUMBER	ORIGIN	INSERTION	ACTION
External oblique	18	*ribs 5–12*		
Internal oblique	23			
Pectoralis minor	13		*coracoid process of scapula*	
Rhomboid major	25			
Rhomboid minor	25	*ligamentum nuchae; spinous processes of C7–T1*		
Teres major	26			*adducts, medially rotates, and extends arm*
Teres minor	26			

QUIZ │ Human Cadaver

Answer the following questions using information from PAL 3.1 as well as other course materials including your textbook, lecture, and lab notes.

I. Check Your Understanding

1. Name one action common to both the deltoid and latissimus dorsi muscles.

   _____

   _____

2. Name a muscle that inserts on the pubic crest.

   _____

3. Which muscle has an origin on the occipital bone and an insertion on the spine of the scapula?

   _____

4. Match the muscles, at left, with their correct origin, at right.

   _____ internal oblique   a. sternum

   _____ external oblique   b. pubic symphysis

   _____ rectus abdominis   c. ribs 5–12

   _____ serratus anterior   d. iliac crest

   _____ pectoralis major   e. ribs 1–8

5. Match the muscles, at left, with their correct action, at right.

   _____ supraspinatus   a. adducts arm

   _____ infraspinatus   b. rotates arm laterally

   _____ teres major   c. depresses scapula

   _____ serratus anterior   d. rotates scapula upwards (laterally)

   _____ pectoralis minor   e. abducts arm

BEYOND PAL 6. Name the four muscles that compose the rotator cuff.

   _____

   _____

   _____

7. What is the name of the dense irregular connective tissue surrounding all skeletal muscles?

   _____

   _____

II. Apply What You Learned

1. Curare is a poison well known for its paralyzing effects. It has historically been used by the indigenous populations of South America in hunting. Arrows or blowgun darts were dipped in the plant-derived curare, which caused relaxation and paralysis of all skeletal muscles. Death usually resulted from asphyxiation due to paralysis of the muscles used for respiration. What would these muscles be?

   _____

   _____

2. Hypotonia, or low muscle tone, is a symptom of many different disorders including Down syndrome. Infants who exhibit hypotonia have difficulty learning to coordinate their muscles, and consequently often learn to roll, sit up, and walk much later than other children. Assume you are a physical therapist working to help a child with low muscle tone learn to sit up. Which key muscle *groups* would you focus on strengthening?

   _____

   _____

   _____

   _____

SELF REVIEW | Anatomical Models

Exercise 3.12 **Superficial and Intermediate Muscles, Anterior View**

GO TO ANATOMICAL MODELS > MUSCULAR SYSTEM > TRUNK > SELF REVIEW > IMAGES 1 AND 2

- *Mouse over the images and locate and label the muscles indicated below. Click on the muscles to hear their pronunciations.*

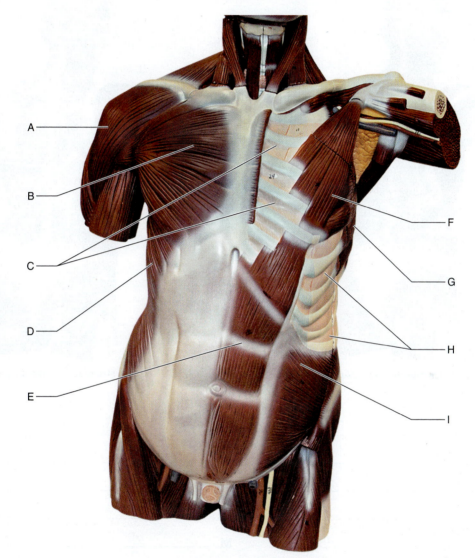

LT-VA16: Life-size muscle torso, 27-part, 3B Scientific®

A. _____

B. _____

C. _____

D. _____

E. _____

F. _____

G. _____

H. _____

I. _____

Exercise 3.13 **Superficial, Intermediate, and Deep Muscles, Posterior View**

GO TO ANATOMICAL MODELS > MUSCULAR SYSTEM > TRUNK > SELF REVIEW > IMAGES 3 AND 4

- *Mouse over the images and locate and label the muscles indicated below. Click on the muscles to hear their pronunciations.*

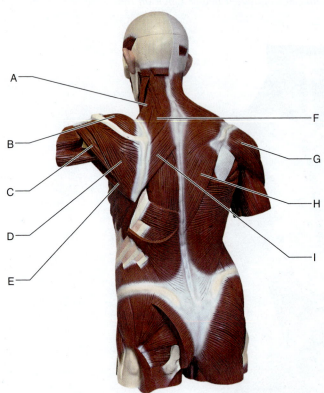

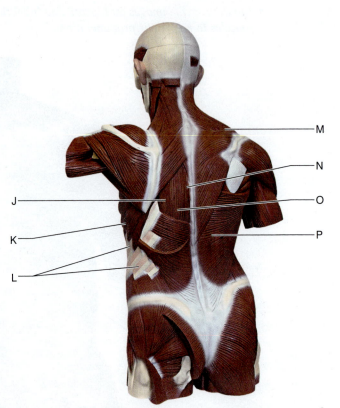

LT-VA16: Life-size muscle torso, 27-part, 3B Scientific®

LT-VA16: Life-size muscle torso, 27-part, 3B Scientific®

A. _____

B. _____

C. _____

D. _____

E. _____

F. _____

G. _____

H. _____

I. _____

J. _____

K. _____

L. _____

M. _____

N. _____

O. _____

P. _____

Animations Questions

GO TO ANATOMICAL MODELS > MUSCULAR SYSTEM > TRUNK > SELF REVIEW

- *Go to the image number next to each muscle listed in the table below.*
- *Locate the muscle and double-click its label to view the animation.*
- *Complete the table below.*

MUSCLE	IMAGE NUMBER	ORIGIN	INSERTION	ACTION
Infraspinatus	3			
Supraspinatus	3	*supraspinous fossa of scapula*		
Latissimus dorsi	7			*extends, adducts, and medially rotates arm*
Levator scapulae	3		*upper 1/3 medial border of scapula*	
Longissimus	4			
Serratus anterior	7			*abducts scapula; laterally rotates scapula (upward); superior fibers elevate and inferior fibers depress scapula*
Trapezius	4			

QUIZ Anatomical Models

Answer the following questions using information from PAL 3.1 as well as other course materials including your textbook, lecture, and lab notes.

I. Check Your Understanding

1. What are the three muscles that compose the erector spinae group?

_____

_____

_____

2. Name two muscles that can rotate the arm medially.

_____

_____

3. List the main muscle(s) used in normal inspiration.

_____

_____

4. Onto which bone(s) does the rhomboid major insert?

_____

_____

5. List the main muscle(s) involved in shrugging your shoulders (elevating your scapulae).

_____

_____

6. Name two muscles that have an origin on the spinous process of at least one thoracic vertebra.

_____

_____

7. Match the muscles, at left, with their correct insertion, at right.

_____ pectoralis minor
_____ deltoid
_____ latissimus dorsi
_____ teres major
_____ pectoralis major

a. coracoid process of scapula

b. lesser tubercle of humerus

c. greater tubercle of humerus

d. deltoid tuberosity of humerus

e. intertubercular groove of humerus

II. Apply What You Learned

1. After falling off her bike, a woman goes to her doctor complaining of shoulder pain. After performing an exam and asking her to try various shoulder movements, the doctor tells her that she has a rotator cuff injury and that she has obvious impairment of abduction and lateral rotational movements of the arm.

a. Which muscle(s) of the rotator cuff could be injured?

_____

_____

_____

b. For each of the rotator cuff muscles, indicate how the muscle contributes to shoulder movement and function.

_____

_____

_____

MUSCLES OF THE UPPER LIMB

SELF REVIEW Human Cadaver

Exercise 3.14 **Intermediate Arm Muscles, Anterior View**

GO TO HUMAN CADAVER > MUSCULAR SYSTEM > UPPER LIMB > SELF REVIEW > IMAGE 16

- *Mouse over the image to locate and label the muscles indicated below. Click on the muscles to hear their pronunciations.*

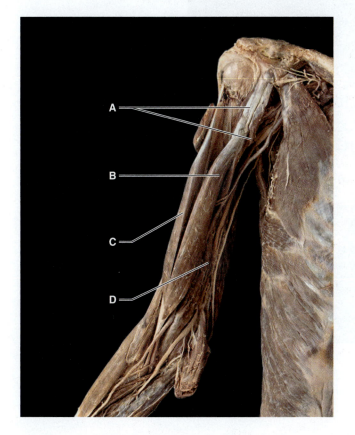

A. _____

B. _____

C. _____

D. _____

Exercise 3.15 **Superficial Arm Muscles, Posterior View**

GO TO ⟩ HUMAN CADAVER > MUSCULAR SYSTEM > UPPER LIMB > SELF REVIEW > IMAGE 19

> • *Mouse over the image to locate and label the muscles indicated below. Click on the muscles to hear their pronunciations.*

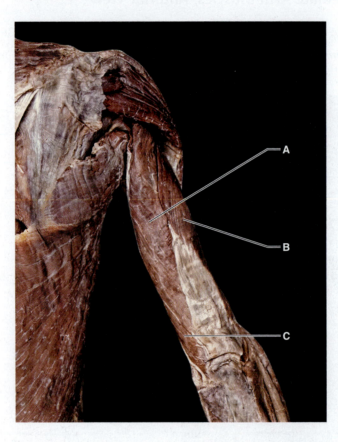

A. _____   C. _____

B. _____

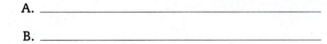

Exercise 3.16   **Superficial and Deep Forearm Muscles, Anterior View**

GO TO HUMAN CADAVER > MUSCULAR SYSTEM > UPPER LIMB > SELF REVIEW > IMAGES 20 AND 22

  • *Mouse over the images and locate and label the muscles indicated below. Click on the muscles to hear their pronunciations.*

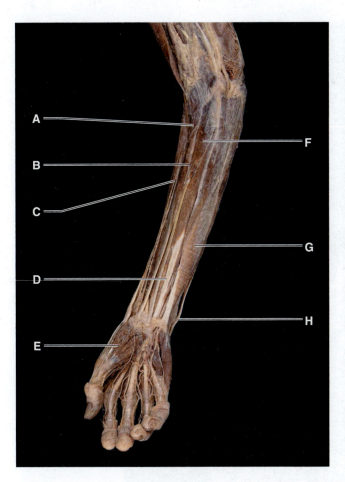
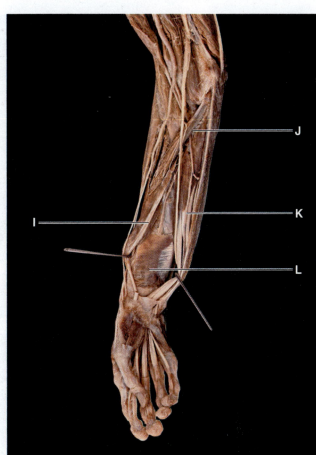

A. _____   G. _____

B. _____   H. _____

C. _____   I. _____

D. _____   J. _____

E. _____   K. _____

F. _____   L. _____

Exercise 3.17 **Superficial Forearm Muscles, Lateral View**

GO TO HUMAN CADAVER > MUSCULAR SYSTEM > UPPER LIMB > SELF REVIEW > IMAGE 23

- *Mouse over the image to locate and label the muscles indicated below. Click on the muscles to hear their pronunciations.*

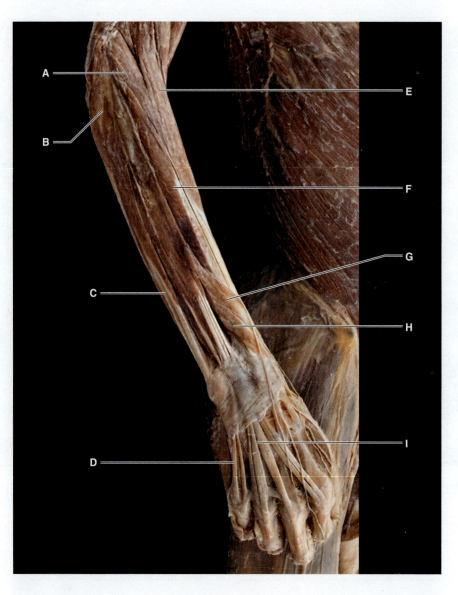

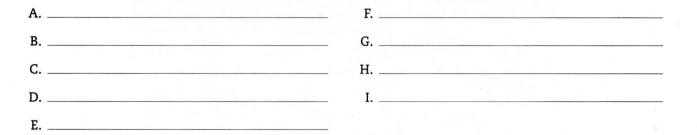

A. _____   F. _____

B. _____   G. _____

C. _____   H. _____

D. _____   I. _____

E. _____

Exercise 3.18 **Intermediate Hand Muscles, Anterior View**

GO TO > HUMAN CADAVER > MUSCULAR SYSTEM > UPPER LIMB > SELF REVIEW > IMAGE 27

> • *Mouse over the image to locate and label the muscles indicated below. Click on the*
> *muscles to hear their pronunciations.*

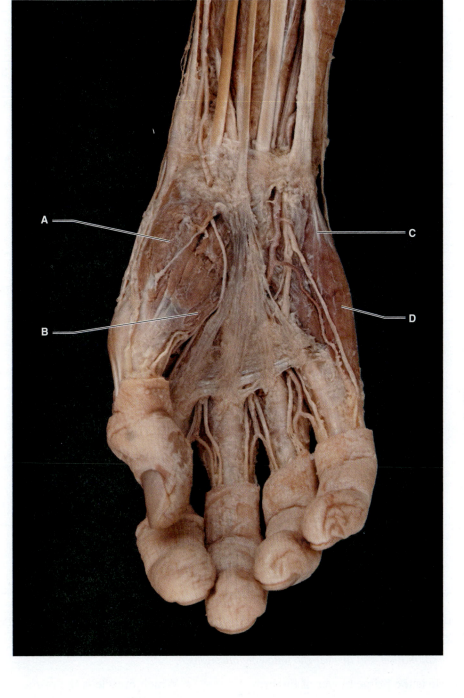

A. _____   C. _____

B. _____   D. _____

Animations Questions

GO TO ▷ HUMAN CADAVER > MUSCULAR SYSTEM > UPPER LIMB > SELF REVIEW

- *Go to the image number next to each muscle listed in the table below.*
- *Locate the muscle and double-click its label to view the animation.*
- *Complete the table below.*

MUSCLE	IMAGE NUMBER	ORIGIN	INSERTION	ACTION
Biceps brachii	15	Long head: Short head:	Long head: Short head:	Long & Short head: *flexes forearm at elbow; supinates forearm*
Brachialis	18	*distal 1/2 of anterior surface of humerus*		
Brachioradialis	5			*flexes forearm at elbow*
Extensor carpi radialis brevis	24		*base of metacarpal 3*	
Extensor carpi radialis longus	24			
Flexor carpi radialis	20		*bases of metacarpals 2–3*	
Deltoid	12			

GO TO ▷ PAL 3.1 HOME > ANIMATIONS (UPPER RIGHT CORNER) > JOINT MOVEMENTS AND ACTIONS OF MUSCLE GROUPS > ELBOW AND FOREARM > ELBOW JOINT 2

- *Play the animation and answer the following questions.*

1. Which muscle is the prime mover of forearm flexion?

_____

2. Which two muscles act as synergists to the prime mover of forearm flexion?

_____

_____

3. Which muscle is the prime mover of forearm extension?

_____

QUIZ | Human Cadaver

Answer the following questions using information from PAL 3.1 as well as other course materials including your textbook, lecture, and lab notes.

I. Check Your Understanding

1. From which structure and bone does the short head of the biceps brachii originate?

_____

_____

2. What is the action of the anconeus muscle?

_____

_____

3. Name a muscle that inserts on the palmar aponeurosis.

_____

4. Which muscle has an origin on the infraglenoid tubercle of the scapula, and an insertion on the olecranon process of the ulna?

_____

5. Match the muscles, at left, with their correct origin, at right.

_____ pronator teres

_____ extensor carpi ulnaris

_____ flexor carpi radialis

_____ abductor pollicis longus

_____ brachioradialis

a. lateral supracondylar ridge of humerus

b. coronoid process of ulna

c. lateral epicondyle of humerus

d. posterior radius

e. medial epicondyle of humerus

6. Match the muscles, at left, with their correct action, at right.

_____ extensor carpi radialis longus

_____ extensor carpi ulnaris

_____ palmaris longus

_____ extensor indicis

_____ biceps brachii

a. extends hand at wrist

b. supinates forearm

c. flexes hand at wrist

d. adducts hand

e. extends digit 2

7. What is the dense regular connective tissue structure that connects muscle to bone?

_____

_____

II. Apply What You Learned

1. The median nerve passes through the carpal tunnel of the wrist on its way to innervate structures of the hand. If there is compression of this nerve by surrounding structures, it can cause significant pain, tingling, and numbness of the hand.

   a. Which muscles have tendons that also pass through the carpal tunnel?

   _____

   _____

   b. What is the major action(s) being performed by these muscles that might contribute to overuse and inflammation problems leading to carpal tunnel?

   _____

   _____

2. In the last 20 years, there has been a dramatic increase in injuries related to the use of computers. In particular, the computer mouse has resulted in many contorted hands and sore forearms due to repetitive muscular movements. Imitate the left to right movement your hand makes while using a mouse (or, if you have your computer mouse handy, perform the real movement). You can probably feel which muscles of the forearm are involved in the abduction/adduction movements at your wrist. What are the major forearm muscles involved in each of these movements?

_____

_____

_____

_____

SELF REVIEW │ Anatomical Models

Exercise 3.19 **Superficial Scapula and Arm Muscles, Anterior View**

GO TO ⟩ ANATOMICAL MODELS > MUSCULAR SYSTEM > UPPER LIMB > SELF REVIEW > IMAGE 1

- *Mouse over the image to locate and label the muscles indicated below. Click on the muscles to hear their pronunciations.*

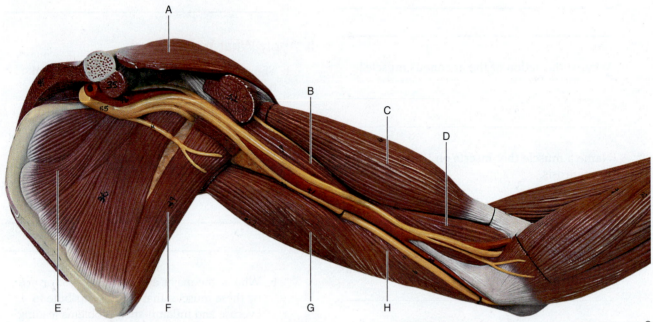

LT-M11: Deluxe muscular arm, 6-part, 3B Scientific®

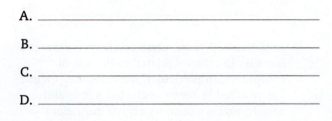

A. _____   E. _____

B. _____   F. _____

C. _____   G. _____

D. _____   H. _____

Exercise 3.20   **Superficial Forearm Muscles, Anterior View**

GO TO ⟩ ANATOMICAL MODELS > MUSCULAR SYSTEM > UPPER LIMB > SELF REVIEW > IMAGE 3

- *Mouse over the image to locate and label the muscles indicated below. Click on the muscles to hear their pronunciations.*

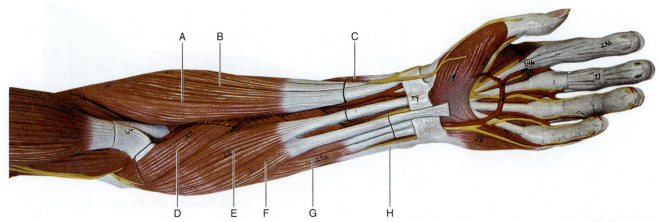

LT-M11: Deluxe muscular arm, 6-part, 3B Scientific®

A. _____   E. _____

B. _____   F. _____

C. _____   G. _____

D. _____   H. _____

Exercise 3.21 **Superficial Forearm and Hand Muscles, Posterior View**

GO TO ANATOMICAL MODELS > MUSCULAR SYSTEM > UPPER LIMB > SELF REVIEW > IMAGE 5

- *Mouse over the image to locate and label the muscles indicated below. Click on the muscles to hear their pronunciations.*

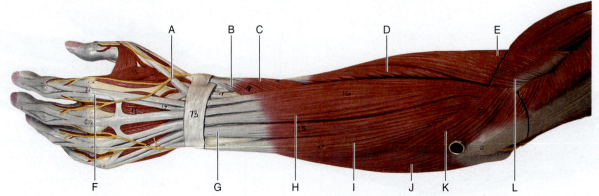

LT-M11: Deluxe muscular arm, 6-part, 3B Scientific®

A. _____

B. _____

C. _____

D. _____

E. _____

F. _____

G. _____

H. _____

I. _____

J. _____

K. _____

L. _____

Exercise 3.22 **Superficial Hand Muscles, Anterior View**

GO TO ANATOMICAL MODELS > MUSCULAR SYSTEM > UPPER LIMB > SELF REVIEW > IMAGE 6

- *Mouse over the image to locate and label the muscles indicated below. Click on the muscles to hear their pronunciations.*

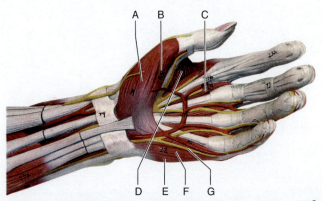

LT-M11: Deluxe muscular arm, 6-part, 3B Scientific®

A. _____

B. _____

C. _____

D. _____

E. _____

F. _____

G. _____

Animations Questions

[GO TO]> ANATOMICAL MODELS > MUSCULAR SYSTEM > UPPER LIMB > SELF REVIEW

- *Go to the image number next to each muscle listed in the table below.*
- *Locate the muscle and double-click its label to view the animation.*
- *Complete the table below.*

MUSCLE	IMAGE NUMBER	ORIGIN	INSERTION	ACTION
Extensor carpi ulnaris	5		base of metacarpal 5 (medial surface)	
Flexor carpi ulnaris	5	*humeral–medial epicondyle of humerus; ulnar–olecranon process and proximal 2/3 of posterior ulna*		
Extensor digitorum	5			*extends all joints of digits 2–5*
Flexor digitorum superficialis	3			
Pronator teres	3	*superior to medial epicondyle of humerus; coronoid process of ulna (medial side)*		
Supinator	4			*supinates forearm*
Triceps brachii	2	Lateral head: Long head: Medial head:	Lateral head: Long head: Medial head:	Lateral head: Long head: Medial head:

QUIZ Anatomical Models

Answer the following questions using information from PAL 3.1 as well as other course materials including your textbook, lecture, and lab notes.

I. Check Your Understanding

1. From which structure and bone does the long head of the triceps brachii originate?

   _____

   _____

2. What is an action of the lumbrical muscles?

   _____

   _____

3. Which heads of the triceps brachii assist with elbow extension?

   _____

   _____

4. Which nerve innervates the extensor carpi radialis longus and extensor carpi radialis brevis?

   _____

   _____

5. Which muscle flexes and abducts the hand at the wrist?

   _____

   _____

6. Match the muscles, at left, with their correct insertion, at right.

   _____ brachioradialis   a. medial shaft of humerus

   _____ brachialis

   _____ triceps brachii   b. radial tuberosity of radius
         (all heads)
                                  c. coronoid process of ulna
   _____ coracobrachialis

   _____ biceps brachii   d. olecranon process of ulna
         (both heads)
                                  e. styloid process of radius

II. Apply What You Learned

1. While trying to remove a jam from a hay bailer, an 18-year-old man's arm was severed about 1/3 of the way above the elbow. During surgery, which muscles needed to be re-attached?

   _____

   _____

   _____

2. The tendon of the palmaris longus muscle is commonly used when a tendon graft is needed in the wrist. In this process, the tendon of the muscle is cut away from its insertion on the palmar aponeurosis and relocated to the site of the ruptured tendon.

   a. Would the removal of the palmaris longus muscle cause the loss of any movement at the wrist? Why or why not?

   _____

   _____

   _____

   _____

   b. Some individuals do not have the option of using the palmaris longus for a tendon graft. Why is this?

   _____

   _____

   _____

MUSCLES OF THE LOWER LIMB

SELF REVIEW | Human Cadaver

Exercise 3.23 **Superficial Thigh Muscles, Anterior View**

GO TO〉 HUMAN CADAVER > MUSCULAR SYSTEM > LOWER LIMB > SELF REVIEW > IMAGE 16

- *Mouse over the image to locate and label the muscles indicated below. Click on the muscles to hear their pronunciations.*

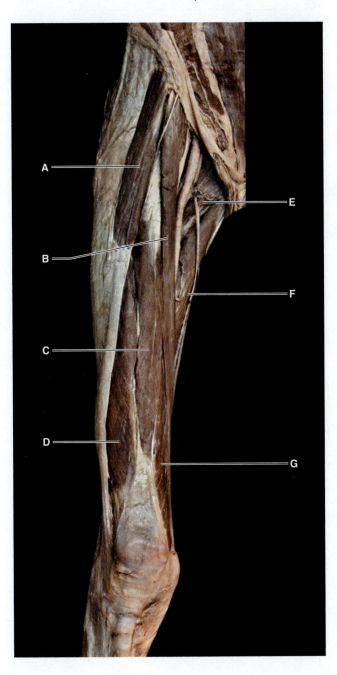

A. _____

B. _____

C. _____

D. _____

E. _____

F. _____

G. _____

Exercise 3.24 **Superficial Thigh Muscles, Posterior View**

GO TO ⟩ HUMAN CADAVER > MUSCULAR SYSTEM > LOWER LIMB > SELF REVIEW > IMAGE 22

- *Mouse over the image to locate and label the muscles indicated below. Click on the muscles to hear their pronunciations.*

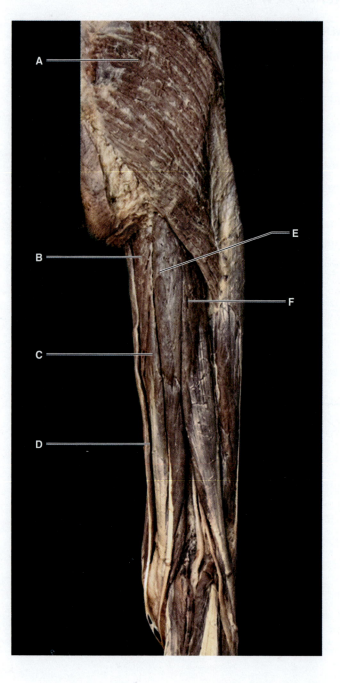

A. _____

B. _____

C. _____

D. _____

E. _____

F. _____

Exercise 3.25   **Deep Thigh Muscles, Posterior View**

GO TO > HUMAN CADAVER > MUSCULAR SYSTEM > LOWER LIMB > SELF REVIEW > IMAGE 24

- *Mouse over the image to locate and label the muscles indicated below. Click on the muscles to hear their pronunciations.*

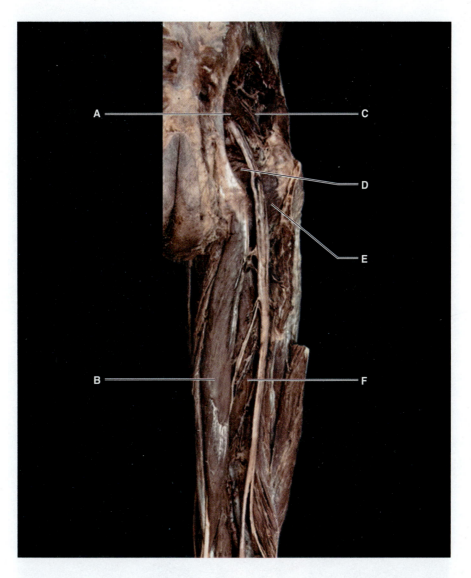

A. _____   D. _____

B. _____   E. _____

C. _____   F. _____

Exercise 3.26   **Intermediate Thigh Muscles, Medial View**

GO TO ▷ HUMAN CADAVER > MUSCULAR SYSTEM > LOWER LIMB > SELF REVIEW > IMAGE 27

- *Mouse over the image to locate and label the muscles indicated below. Click on the muscles to hear their pronunciations.*

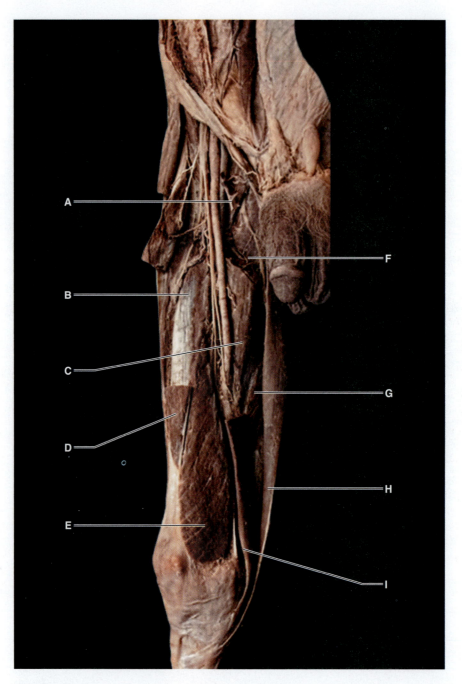

A. _____   F. _____

B. _____   G. _____

C. _____   H. _____

D. _____   I. _____

E. _____

Exercise 3.27   **Superficial Leg Muscles, Lateral View**

GO TO HUMAN CADAVER > MUSCULAR SYSTEM > LOWER LIMB > SELF REVIEW > IMAGE 48

- *Mouse over the image to locate and label the muscles indicated below. Click on the muscles to hear their pronunciations.*

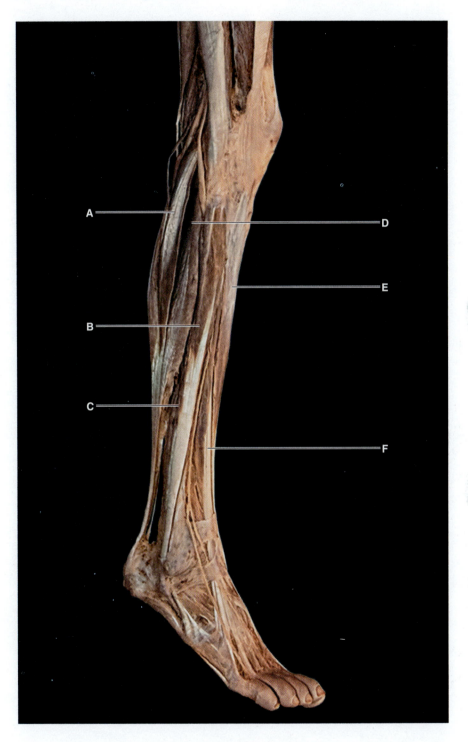

A. _____   D. _____

B. _____   E. _____

C. _____   F. _____

Exercise 3.28   **Deep Leg Muscles, Posterior View**

GO TO   HUMAN CADAVER > MUSCULAR SYSTEM > LOWER LIMB > SELF REVIEW > IMAGE 52

  • *Mouse over the image to locate and label the muscles indicated below. Click on the muscles to hear their pronunciations.*

A. _____   C. _____

B. _____   D. _____

Animations Questions

GO TO ⟩ HUMAN CADAVER > MUSCULAR SYSTEM > LOWER LIMB > SELF REVIEW

- *Go to the image number next to each muscle listed in the table below.*
- *Locate the muscle and double-click its label to view the animation.*
- *Complete the table below.*

MUSCLE	IMAGE NUMBER	ORIGIN	INSERTION	ACTION
Biceps femoris	22	Long head: *ischial tuberosity* Short head:	Long head: Short head:	Long head: Short head:
Extensor digitorum longus	48		*divides into four tendons to the middle and distal phalanges of digits 2–5*	
Flexor digitorum longus	56			
Gluteus maximus	21	*posterior surfaces of ilium, iliac crest, inferior sacrum, coccyx*		
Semimembranosus	22			
Semitendinosus	22			*extends thigh at hip; flexes leg at knee*
Tibialis anterior	46			
Tibialis posterior	56		*navicular (medial surface); three cuneiforms (plantar surface); bases of metatarsals 2–4*	

GO TO ⟩ HOME PAGE > ANIMATIONS (UPPER RIGHT CORNER) > JOINT MOVEMENTS AND ACTIONS
OF MUSCLE GROUPS > HIP AND FEMUR > HIP JOINT 5

• *Play the animation and answer the following questions.*

1. Which two muscles are the prime movers of hip flexion?

_____

_____

2. Which two muscles are responsible for hip extension?

_____

_____

3. Which muscle is the prime mover of hip adduction?

_____

QUIZ ⟩ Human Cadaver

Answer the following questions using information from PAL 3.1 as well as other course materials including your textbook, lecture, and lab notes.

I. Check Your Understanding

1. Name the four muscle heads that compose the quadriceps femoris.

_____

_____

_____

_____

2. Which muscle is the prime mover in dorsiflexion of the foot?

_____

3. Name two muscles that insert onto the calcaneus bone via the calcaneal tendon.

_____

_____

4. Name a muscle that has origins on both the fibula and the lateral condyle of the tibia.

_____

5. Name two muscles that have an origin on the ischial tuberosity and flex the leg at the knee?

_____

_____

6. Match the muscles below, at left, with their correct insertion, at right.

_____ semimembranosus

_____ sartorius

_____ adductor magnus

_____ biceps femoris

_____ tibialis posterior

a. proximal tibia
b. linea aspera of femur
c. head of fibula
d. medial condyle of tibia
e. navicular

7. Match the muscles below, at left, with their correct action, at right.

_____ adductor longus

_____ gracilis

_____ gluteus medius

_____ biceps femoris, long head

_____ vastus lateralis

a. extends thigh at hip
b. extends leg at knee
c. flexes leg at knee
d. flexes thigh at hip
e. abducts thigh

II. Apply What You Learned

1. Compression or paralysis of the deep fibular nerve can result in a condition called foot drop. Motor innervation of which muscles would be impacted, and which major movement of the ankle would be compromised?

_____

_____

_____

_____

_____

2. Shin splints are a common and painful condition of the lower limb. The pain is typically felt along the anterior or lateral surface of the leg. The causes can be varied. Three of the most common causes of shin splits are listed below. Explain why each of these would cause the pain associated with the condition.

a. Overuse of muscles

_____

_____

_____

b. Stress fracture (hairline breaks) in bone

_____

_____

_____

c. Overpronation of feet

_____

_____

_____

SELF REVIEW | Anatomical Models

Exercise 3.29 **Superficial Thigh Muscles, Anterior View**

GO TO > ANATOMICAL MODELS > MUSCULAR SYSTEM > LOWER LIMB > SELF REVIEW IMAGE 5

- *Mouse over the image to locate and label the muscles indicated below. Click on the muscles to hear their pronunciations.*

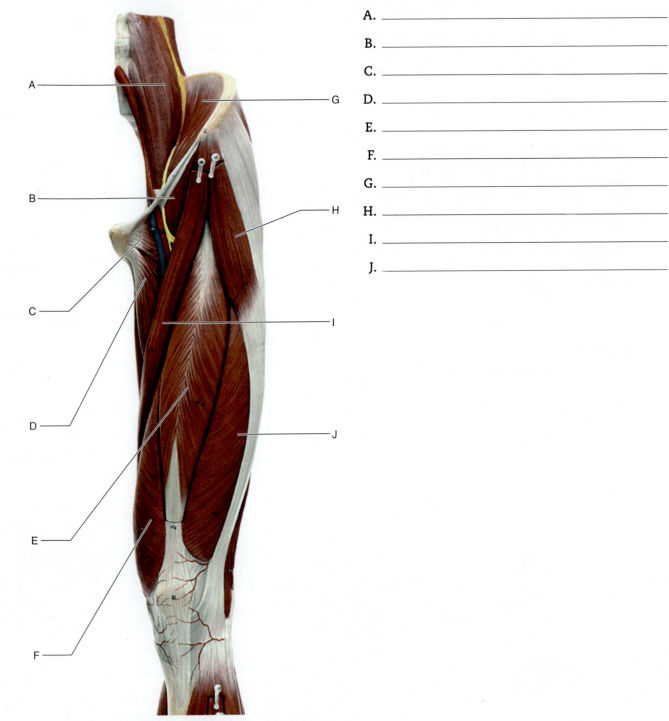

A. _____

B. _____

C. _____

D. _____

E. _____

F. _____

G. _____

H. _____

I. _____

J. _____

Exercise 3.30   Superficial Thigh Muscles, Posterior View

GO TO ⟩ ANATOMICAL MODELS > MUSCULAR SYSTEM > LOWER LIMB > SELF REVIEW > IMAGE 7

- *Mouse over the image to locate and label the muscles indicated below. Click on the muscles to hear their pronunciations.*

A. _____

B. _____

C. _____

D. _____

E. _____

F. _____

G. _____

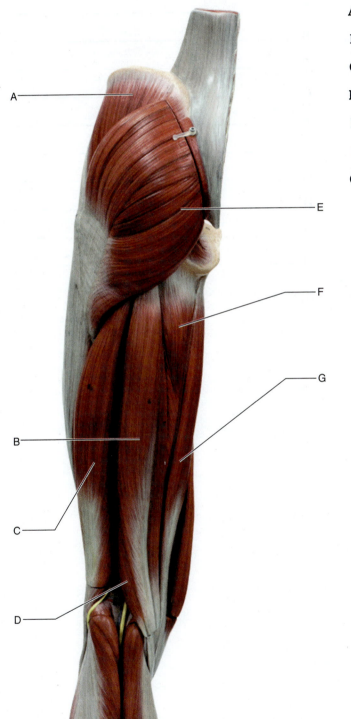

Exercise 3.31   **Deep Thigh Muscles, Posterior View**

GO TO〉 ANATOMICAL MODELS > MUSCULAR SYSTEM > LOWER LIMB > SELF REVIEW > IMAGE 8

- *Mouse over the image to locate and label the muscles indicated below. Click on the muscles to hear their pronunciations.*

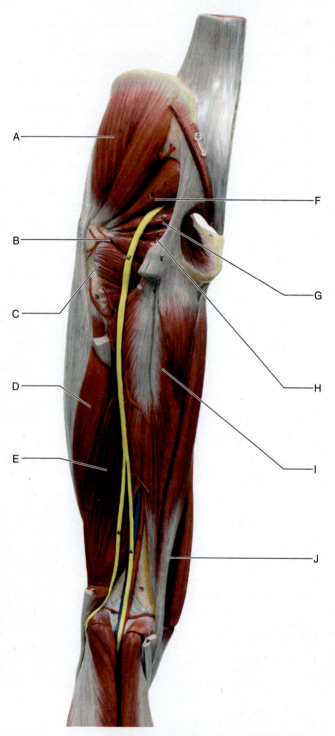

A. _____

B. _____

C. _____

D. _____

E. _____

F. _____

G. _____

H. _____

I. _____

J. _____

Exercise 3.32   **Superficial Leg and Foot Muscles, Lateral View**

GO TO ⟩ ANATOMICAL MODELS > MUSCULAR SYSTEM > LOWER LIMB > SELF REVIEW > IMAGE 11

- *Mouse over the image to locate and label the muscles indicated below. Click on the muscles to hear their pronunciations.*

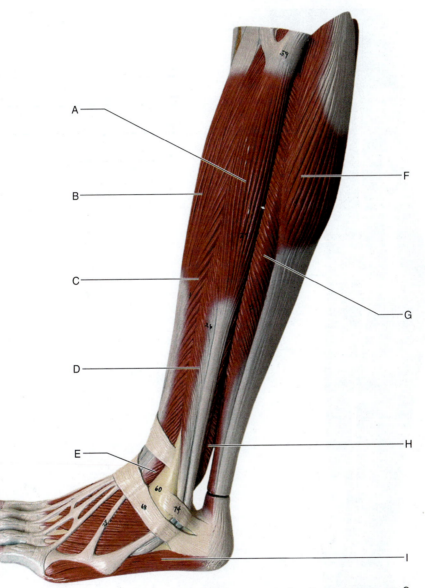

LT-M21: Muscular leg, 7-part, 3B Scientific®

A. _____   F. _____

B. _____   G. _____

C. _____   H. _____

D. _____   I. _____

E. _____

Exercise 3.33 **Deep Leg Muscles, Posterior View**

GO TO > ANATOMICAL MODELS > MUSCULAR SYSTEM > LOWER LIMB > SELF REVIEW > IMAGE 17

- *Mouse over the image to locate and label the muscles indicated below. Click on the muscles to hear their pronunciations.*

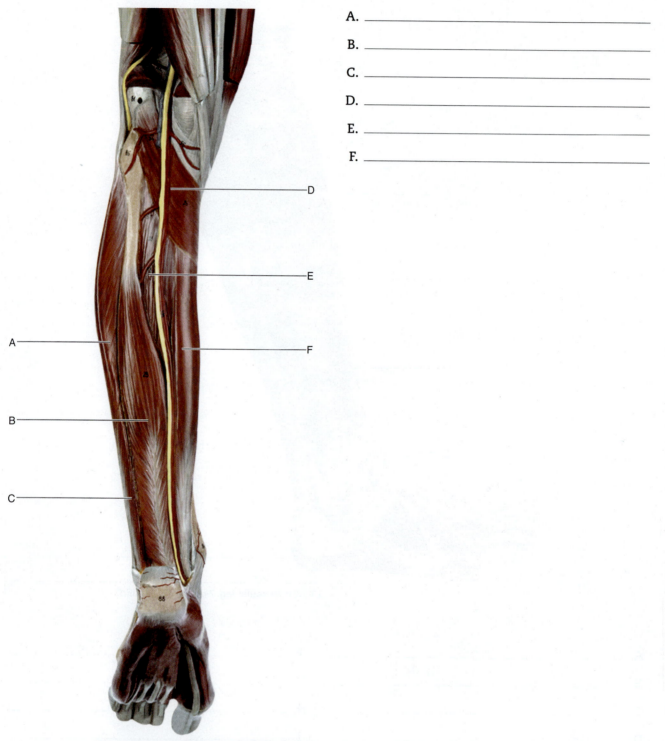

A. _____

B. _____

C. _____

D. _____

E. _____

F. _____

Animations Questions

GO TO ANATOMICAL MODELS > MUSCULAR SYSTEM > LOWER LIMB > SELF REVIEW

- *Go to the image number next to each muscle listed in the table below.*
- *Locate the muscle and double-click its label to view the animation.*
- *Complete the table below.*

MUSCLE	IMAGE NUMBER	ORIGIN	INSERTION	ACTION
Extensor hallucis longus	9			
Flexor hallucis longus	13	*posterior fibula; interosseous membrane*		
Gastrocnemius	11			*plantar flexes foot at ankle; assists in flexion of leg at knee*
Soleus	11			
Rectus femoris	1		*patella and tibial tuberosity via patellar ligament*	
Vastus intermedius	2			
Vastus lateralis	2			
Vastus medialis	2			*extends leg at knee*
Sartorius	1			*flexes thigh at hip; laterally rotates and abducts thigh; flexes knee*

GO TO > HOME PAGE > ANIMATIONS (UPPER RIGHT CORNER) > JOINT MOVEMENTS AND ACTIONS OF MUSCLE GROUPS > ANKLE AND FOOT > ANKLE FOOT 4A

- *Play the animation and answer the following questions.*

1. Which three muscles make up the deep posterior compartment of the leg?

   _____

   _____

   _____

2. Which muscles of the deep posterior compartment of the leg have an origin on the fibula?

   _____

   _____

3. Which muscle of the deep posterior compartment of the leg inserts on the distal phalanx of the hallux (digit 1)?

   _____

   _____

QUIZ | Anatomical Models

Answer the following questions using information from PAL 3.1 as well as other course materials including your textbook, lecture, and lab notes.

I. Check Your Understanding

1. Name the two muscles that form the iliopsoas.

   _____

   _____

2. Which gluteal muscle laterally rotates the thigh?

   _____

3. Name the muscle that inserts at the base of metatarsal 5 and assists in dorsiflexion of the foot.

   _____

4. Name a thigh adductor that has an origin on the inferior ramus of the pubis.

   _____

5. Which nerve plexus innervates the quadriceps femoris?

   _____

   _____

6. Match the muscles, at left, with their correct action, at right.

   _____ fibularis tertius

   _____ tibialis posterior

   _____ extensor digitorum

   _____ fibularis brevis

   _____ tibialis anterior

   a. inverts and plantar flexes foot

   b. inverts and dorsiflexes foot

   c. extends all joints of digits 2–4

   d. everts and dorsiflexes foot

   e. everts and plantar flexes foot

II. Apply What You Learned

1. A ruptured Achilles tendon is a common sports injury. It is actually the calcaneal tendon that is completely or partially torn in this injury.

   a. Which muscles would lose their attachment to the calcaneus if the tendon completely ruptured?

   _____

b. What type of tissue composing the tendon would be torn?

_____

c. Which action of the ankle would be affected?

_____

_____

2. Sprains and strains are common injuries. These terms are commonly used interchangeably, but medically refer to different types of injuries.

a. What is a sprain?

_____

_____

b. What is a strain?

_____

_____

c. How do they differ from one another in terms of tissue types damaged?

_____

_____

_____

d. Which would you expect to take longer to heal—a sprain or a strain—and why?

_____

_____

_____

TISSUES OF THE MUSCULAR SYSTEM

SELF REVIEW | Histology

Exercise 3.34 **Skeletal Muscle, Longitudinal Section 400x**

GO TO HISTOLOGY > MUSCULAR SYSTEM > SELF REVIEW > IMAGE 3

• *Mouse over the image to locate and label the structures indicated below. Click on the structures to hear their pronunciations.*

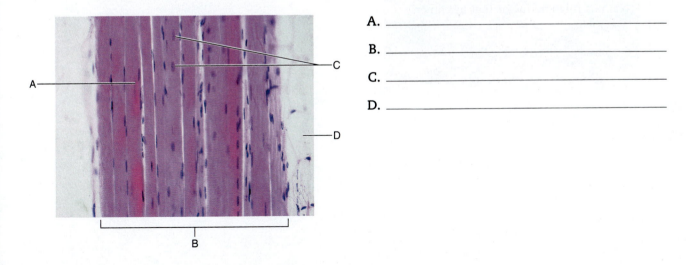

A. _____

B. _____

C. _____

D. _____

Exercise 3.35 **Skeletal Muscle, Cross Section 400x**

GO TO > HISTOLOGY > MUSCULAR SYSTEM > SELF REVIEW > IMAGE 4

• *Mouse over the image to locate and label the structures indicated below. Click on the structures to hear their pronunciations.*

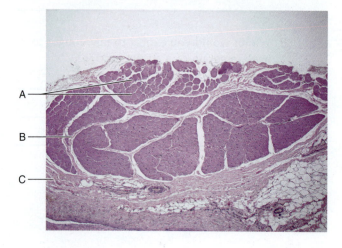

A. _____

B. _____

C. _____

Exercise 3.36 **Neuromuscular Junction 1000x**

GO TO > HISTOLOGY > MUSCULAR SYSTEM > SELF REVIEW > IMAGE 12

• *Mouse over the image to locate and label the structures indicated below. Click on the structures to hear their pronunciations.*

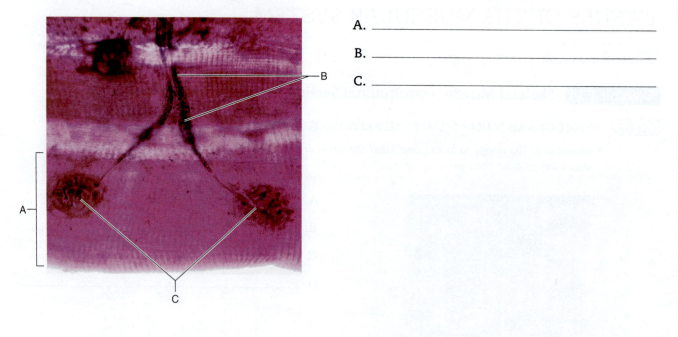

A. _____

B. _____

C. _____

Exercise 3.37   Cardiac Muscle, Longitudinal Section 1000x

GO TO ⟩ HISTOLOGY > MUSCULAR SYSTEM > SELF REVIEW > IMAGE 15

- *Mouse over the image to locate and label the structures indicated below. Click on the structures to hear their pronunciations.*

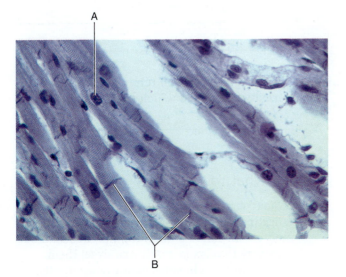

A. _____

B. _____

Exercise 3.38   Smooth Muscle, Duodenum, Cross Section 400x

GO TO ⟩ HISTOLOGY > MUSCULAR SYSTEM > SELF REVIEW > IMAGE 18

- *Mouse over the image to locate and label the structures indicated below. Click on the structures to hear their pronunciations.*

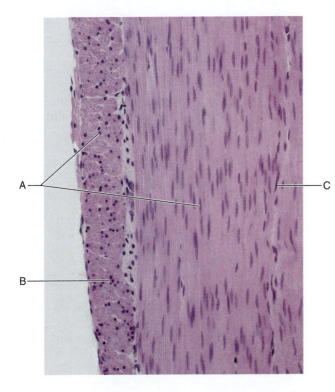

A. _____

B. (layer) _____

C. (layer) _____

QUIZ ⟩ Histology

Answer the following questions using information from PAL 3.1 as well as other course materials including your textbook, lecture, and lab notes.

I. Check Your Understanding

1. a. What type of muscle tissue contains intercalated discs?

_____

b. What is the function of intercalated discs?

_____

_____

2. What is the name of the portion of the sarcolemma that is in contact with the motor axon terminal?

_____

3. What is the name of the synapse between a motor axon terminal and a motor end plate?

_____

4. What is the term used for a bundle of skeletal muscle fibers?

_____

BEYOND **5.** What are the basic functions of skeletal muscle?
PAL

_____

_____

_____

6. Cardiac muscle tissue composes which layer of the heart wall?

_____

_____

7. The sarcoplasmic reticulum of skeletal and cardiac muscle stores what critical ion?

_____

_____

8. Crossbridge formation refers to the bond formation between which two structures?

_____

_____

II. Apply What You Learned

1. Botox© is used in a procedure that involves injecting very small amounts of a bacterial neurotoxin called botulinum into muscle tissue as a way to relax (actually, paralyze) them. The neurotoxin invades motor neuron cells and releases an enzyme that inhibits a muscle contraction.

a. If neurons are the cells targeted by the neurotoxin, why is the muscle function impacted?

_____

_____

_____

_____

_____

b. What is the term used for the location where motor neurons and muscle cells communicate?

_____

_____

2. Compare and contrast the microscopic
   structure, location in the body, and nervous
   control of the three types of muscle tissue.

	MICROSCOPIC STRUCTURE	LOCATION IN BODY	NERVOUS CONTROL
Skeletal muscle tissue			
Cardiac muscle tissue			
Smooth muscle tissue			

LAB PRACTICAL   Muscular System

1. Identify the muscles.

   _____

2. Identify the muscle.

   _____

3. Identify the muscle.

   _____

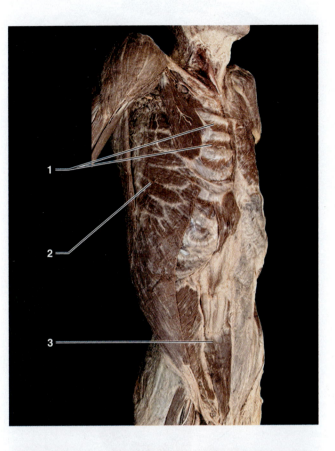

4. Identify the muscle.

   _____

5. Identify the muscle.

   _____

6. Identify the muscle.

   _____

7. As a group, what are these three muscles above known as?

   _____

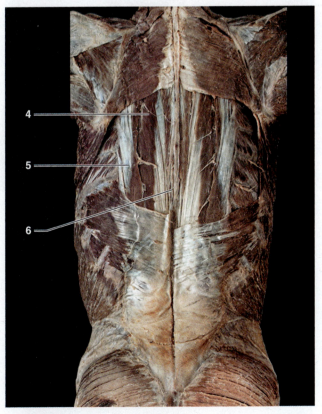

LAB PRACTICAL *continues*

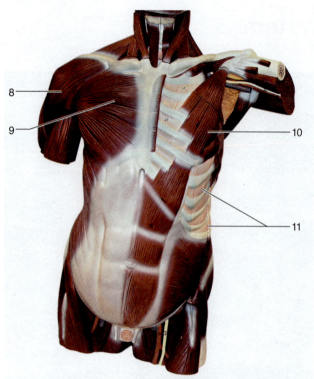

LT-VA16: Life-size muscle torso, 27-part, 3B Scientific®

8. Identify the muscle.

_____

9. Identify the muscle.

_____

10. Onto which bone does this muscle insert?

_____

11. Identify the muscles.

_____

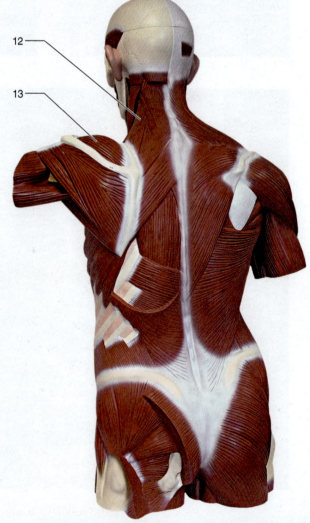

LT-VA16: Life-size muscle torso, 27-part, 3B Scientific®

12. Onto which bone does this muscle insert?

_____

13. From which structure does this muscle originate?

_____

14. Identify the muscle.

_____

15. Identify the muscle.

_____

16. From which bone does this muscle primarily originate?

_____

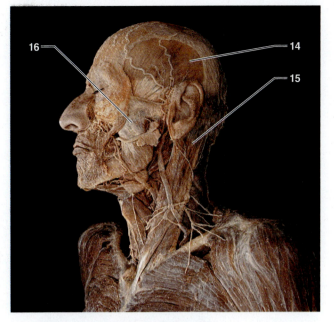

17. Identify the muscle.

_____

18. Identify the muscle.

_____

19. Identify the muscle.

_____

LAB PRACTICAL *continues*

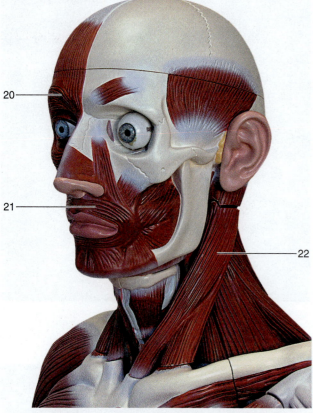

LT-VA16: Life-size muscle torso, 27-part, 3B Scientific®

20. Identify the muscle.

_____

21. Identify the muscle.

_____

22. Onto which structure does this muscle insert?

_____

23. Identify the muscle.

_____

24. What is the action of this muscle?

_____

25. Identify the muscle.

_____

26. Identify the muscle.

_____

27. What action(s) does this muscle have on digit 5?

_____

28. Identify the muscle.

_____

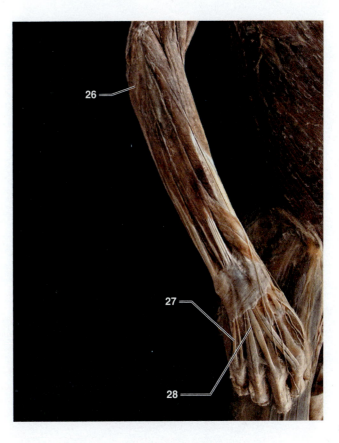

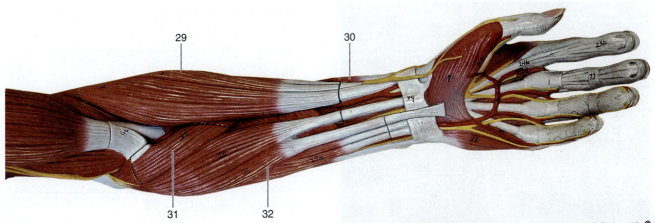

LT-M11: Deluxe muscular arm 6-part, 3B Scientific®

29. Identify the muscle.

_____

30. Identify the muscle.

_____

31. What action does this muscle have?

_____

32. Identify the muscle.

_____

LAB PRACTICAL _continues_

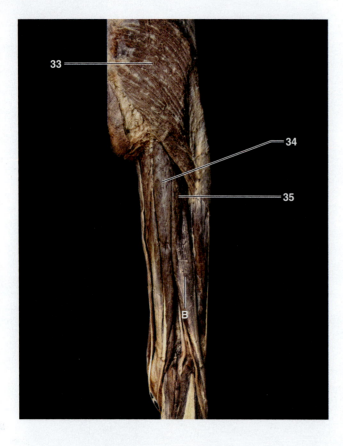

33. Identify the muscle.

_____

34. Identify the muscle.

_____

35. From which structure does this muscle originate?

_____

36. Identify the muscle.

_____

37. What is the action of this muscle?

_____

38. Onto which tarsal bone(s) does this muscle insert?

_____

39. Identify the muscle.

_____

40. From which structure does this muscle originate?

_____

41. Identify the muscle.

_____

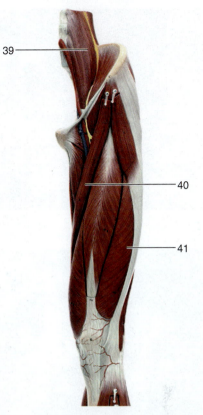

42. Identify the muscle.

_____

43. Onto which bone(s) does this muscle insert?

_____

44. Identify the muscle.

_____

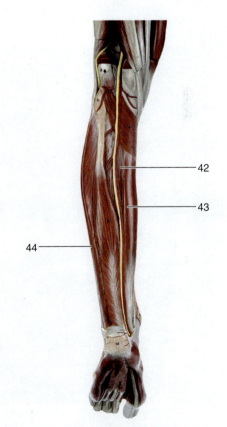

LAB PRACTICAL *continues*

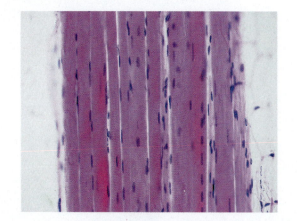

45. Identify the tissue type on the left.

_____

46. Is the tissue type on the left under voluntary or involuntary control?

_____

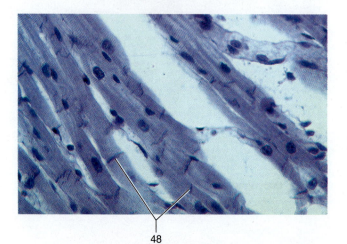

48

47. Identify the tissue type on the left.

_____

48. Identify the structures.

_____

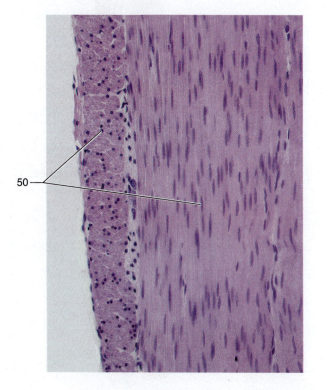

50

49. Identify the tissue type on the left.

_____

50. Identify the structures.

_____

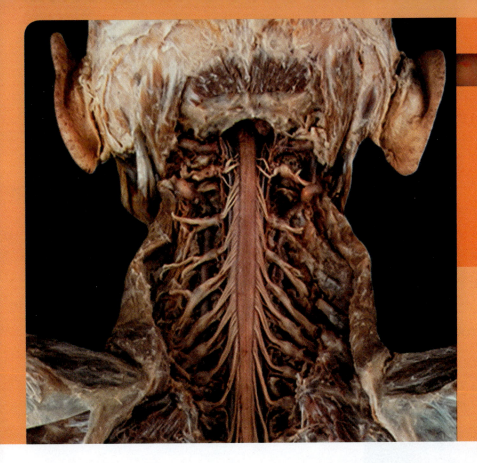

The Nervous System

STUDENT OBJECTIVES

THE CENTRAL NERVOUS SYSTEM

1. List the basic divisions of the nervous system.
2. Define central nervous system and list its components.
3. Name the major regions of the adult brain.
4. Name and locate the ventricles of the brain.
5. List the major lobes and fissures of the brain.
6. Differentiate between commissural, association, and projection fibers.
7. Describe the diencephalon, and name its subdivisions.
8. Identify the three main regions of the brain stem. List the structures located in each region.
9. Describe the structure and function of the cerebellum.
10. Identify and describe the meninges of the brain and spinal cord.
11. Name and locate the dural septa.
12. Describe the gross anatomy of the spinal cord, including the arrangement of its gray and white matter.

THE PERIPHERAL NERVOUS SYSTEM

13. Define peripheral nervous system and list its components.
14. Name the twelve pairs of cranial nerves. Indicate the structures innervated by each pair.
15. Describe the formation of a spinal nerve and the distribution of its rami.
16. Define plexus. Name the major plexuses. Identify the peripheral nerves arising from each plexus.

THE AUTONOMIC NERVOUS SYSTEM

17. Name the organs and structures of the autonomic nervous system.
18. Compare and contrast the anatomy of the parasympathetic and sympathetic divisions of the autonomic nervous system.

THE SPECIAL SENSES

19. Identify the six extrinsic eye muscles, the three eye layers, and the lens.
20. List the structures located within the outer, middle, and inner ear.

21. Identify the different papillae of the tongue.
22. Identify the different structures associated with the sense of smell.

TISSUES OF THE NERVOUS SYSTEM

23. Describe the structural features of a neuron.
24. Describe the general structure of a nerve.

25. Describe the microscopic anatomy of the brain and spinal cord, including the arrangement of the gray and white matter.
26. Identify the three layers of the retina. Name the predominant cell type located within each layer.
27. Describe the microscopic anatomy of the cochlea.
28. Describe the receptor for taste.

THE CENTRAL NERVOUS SYSTEM

SELF REVIEW | Human Cadaver

Exercise 4.1 | Spinal Cord Structures

GO TO > HUMAN CADAVER > NERVOUS SYSTEM > CENTRAL NERVOUS SYSTEM > SELF REVIEW > IMAGES 4 AND 16

- *Mouse over the images to locate and label the structures indicated below. Click on the structures to hear their pronunciations.*

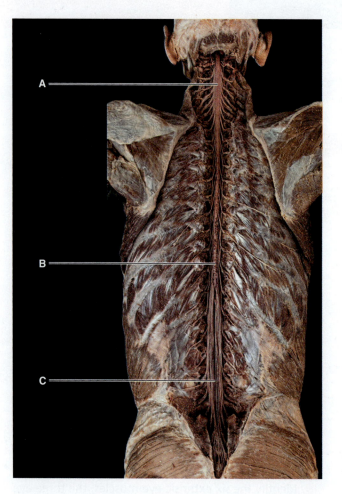

A. _____

B. _____

C. _____

D. _____   F. _____

E. _____   G. _____

Exercise 4.2 **Cervical Enlargement of Spinal Cord**

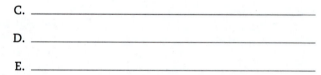

> • *Mouse over the images to locate and label the structures indicated below. Click on the structures to hear their pronunciations.*

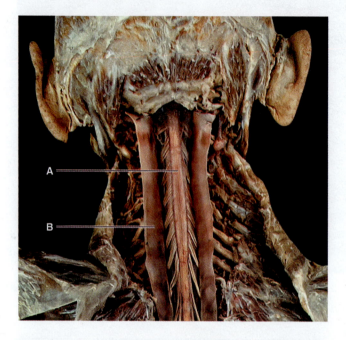

A. (covering)_____

B. (covering)_____

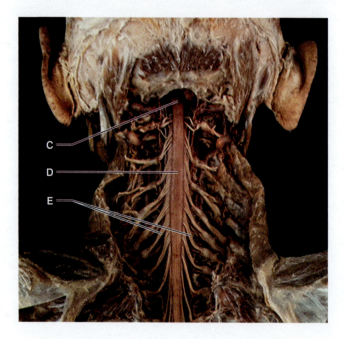

C. _____

D. _____

E. _____

Lobes and Sulci of the Cerebral Cortex

GO TO HUMAN CADAVER > NERVOUS SYSTEM > CENTRAL NERVOUS SYSTEM > SELF REVIEW > IMAGES 21 AND 23

• *Mouse over the images to locate and label the structures indicated below. Click on the structures to hear their pronunciations.*

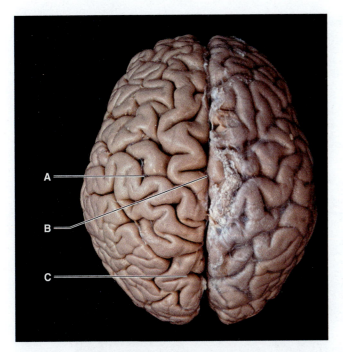

A. (space) _____

B. (space) _____

C. (space) _____

D. (lobe) _____

E. (space) _____

F. (lobe) _____

G. _____

H. (lobe)_____

I. (lobe)_____

Exercise 4.4 Brain Structures, Inferior View

GO TO HUMAN CADAVER > NERVOUS SYSTEM > CENTRAL NERVOUS SYSTEM > SELF REVIEW > IMAGES 25 AND 26

- *Mouse over the images to locate and label the structures indicated below. Click on the structures to hear their pronunciations.*

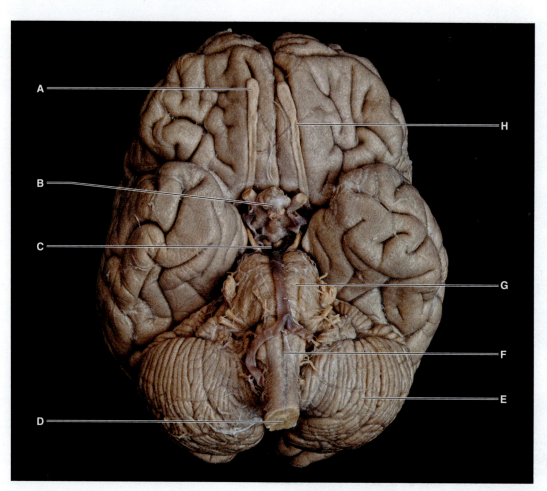

A. _____   E. _____

B. _____   F. _____

C. _____   G. _____

D. _____   H. _____

Exercise 4.5   **Diencephalon, Midbrain, and Cerebellum**

GO TO〉 HUMAN CADAVER > NERVOUS SYSTEM > CENTRAL NERVOUS SYSTEM > SELF REVIEW > IMAGE 29

• *Mouse over the image to locate and label the structures indicated below. Click on the structures to hear their pronunciations.*

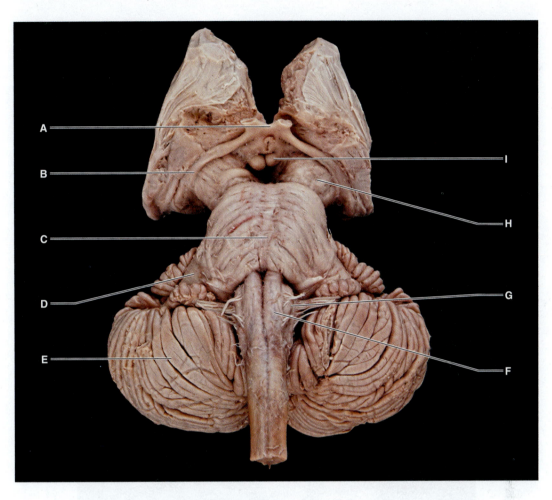

A. _____   F. (specific structure) _____

B. _____   G. (specific structure) _____

C. _____   H. _____

D. _____   I. _____

E. _____

Exercise 4.6 **Brain and Spinal Cord in Bisected Head**

GO TO › HUMAN CADAVER > NERVOUS SYSTEM > CENTRAL NERVOUS SYSTEM > SELF REVIEW > IMAGE 31

• *Mouse over the image to locate and label the structures indicated below. Click on the structures to hear their pronunciations.*

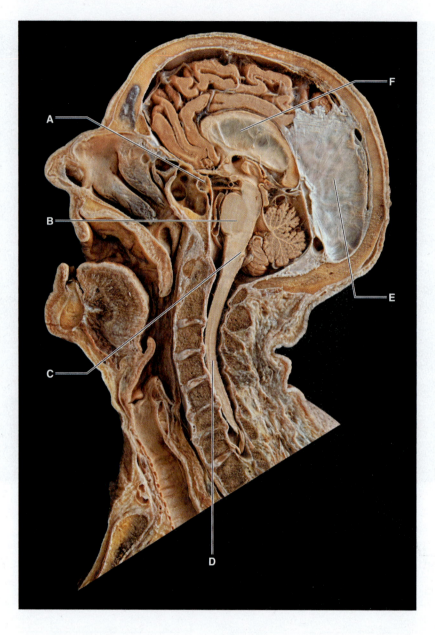

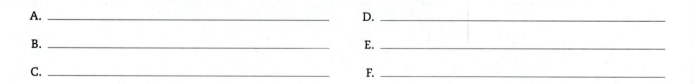

A. _____   D. _____

B. _____   E. _____

C. _____   F. _____

Exercise 4.7 Brain Structures in Bisected Head, Midsagittal View

GO TO HUMAN CADAVER > NERVOUS SYSTEM > CENTRAL NERVOUS SYSTEM > SELF REVIEW > IMAGE 32

- *Mouse over the image to locate and label the structures indicated below. Click on the structures to hear their pronunciations.*

A. _____   D. _____

B. _____   E. _____

C. _____   F. _____

Exercise 4.8   **Midsagittal Brain with Dura Mater**

GO TO HUMAN CADAVER > NERVOUS SYSTEM > CENTRAL NERVOUS SYSTEM > SELF REVIEW > IMAGES 33 AND 34

- *Mouse over the images to locate and label the structures indicated below. Click on the structures to hear their pronunciations.*

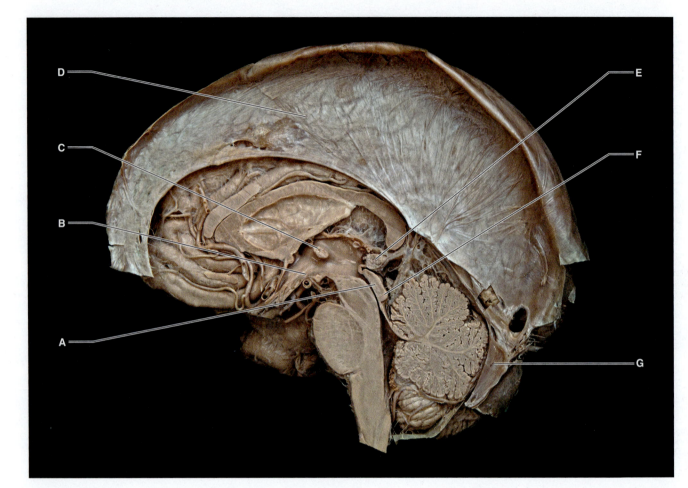

A. (specific structure) _____

B. _____

C. (specific structure) _____

D. _____

E. _____

F. (specific structure) _____

G. _____

Exercise 4.9   Ventricles of the Brain

GO TO ▷ HUMAN CADAVER > NERVOUS SYSTEM > CENTRAL NERVOUS SYSTEM > SELF REVIEW > IMAGE 36

- *Mouse over the image to locate and label the spaces indicated below. Click on the structures to hear their pronunciations.*

A. _____   D. _____

B. _____   E. _____

C. _____   F. _____

Exercise 4.10   Fiber Tracts of the Brain

GO TO HUMAN CADAVER > NERVOUS SYSTEM > CENTRAL NERVOUS SYSTEM > SELF REVIEW > IMAGE 37

- *Mouse over the image to locate and label the structures associated with fiber tracts. Click on the structures to hear their pronunciations.*

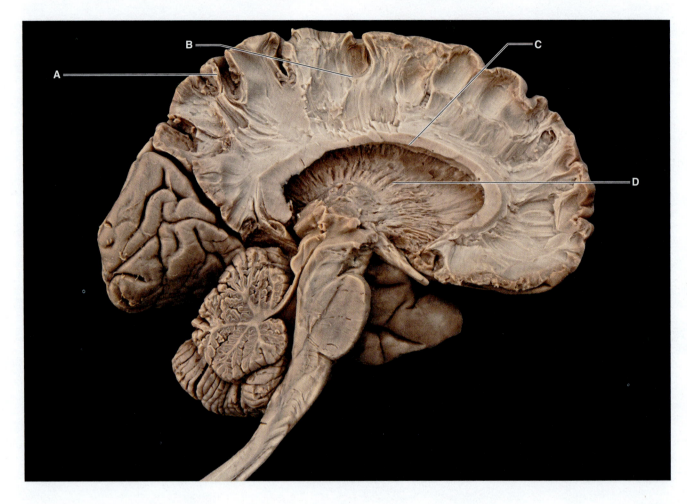

A. _____   C. _____

B. _____   D. _____

Exercise 4.11 **Brain, Anterior Coronal Section**

GO TO HUMAN CADAVER > NERVOUS SYSTEM > CENTRAL NERVOUS SYSTEM > SELF REVIEW > IMAGES 38 AND 39

- *Mouse over the images to locate and label the structures and spaces indicated below. Click on the structures to hear their pronunciations.*

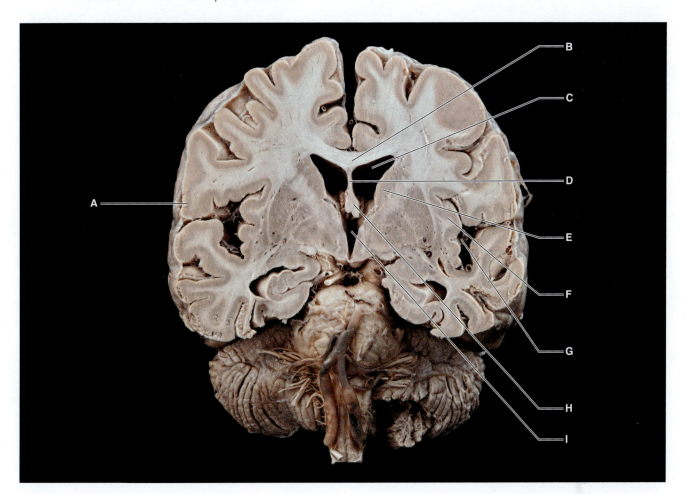

A. _____   F. _____

B. _____   G. (space) _____

C. (space) _____   H. _____

D. _____   I. (space) _____

E. _____

Exercise 4.12 Brain, Transverse Section

GO TO ⟩ HUMAN CADAVER > NERVOUS SYSTEM > CENTRAL NERVOUS SYSTEM > SELF REVIEW > IMAGE 40

- *Mouse over the image to locate and label the structures indicated below. Click on the structures to hear their pronunciations.*

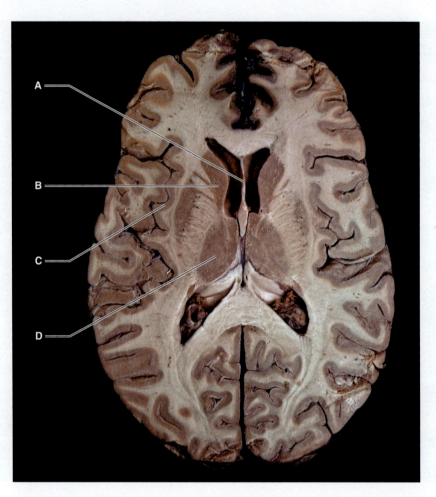

A. _____   C. _____

B. _____   D. _____

Exercise 4.13   **Dura Mater and Dural Sinuses**

GO TO HUMAN CADAVER > NERVOUS SYSTEM > CENTRAL NERVOUS SYSTEM > SELF REVIEW > IMAGES 42 AND 43

> • *Mouse over the images to locate and label the structures indicated below. Click on the structures to hear their pronunciations.*

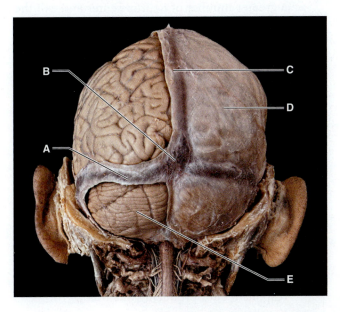

A. _____

B. _____

C. _____

D. _____

E. _____

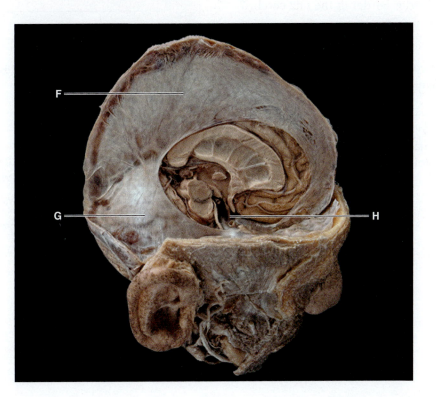

F. _____   H. _____

G. _____

QUIZ ⌉ Human Cadaver

Answer the following questions using information from PAL 3.1 as well as other course materials including your textbook, lecture, and lab notes.

I. Check Your Understanding

1. What are the three meningeal layers of the brain, in order from deepest (closest to the CNS tissue) to most superficial?

_____

_____

_____

2. The central sulcus separates which two cerebral lobes?

_____

_____

3. The dura mater contains several dural sinuses that transport venous blood away from the brain. Which of these dural sinuses is located along the superior border of the falx cerebri?

_____

4. What fibers connect areas within a single cerebral hemisphere?

_____

5. What is the name of the bundle of spinal nerve roots found at the caudal end of the spinal cord?

_____

_____

6. The spinal cord has an enlarged region where a greater number of neurons enter and exit the spinal cord to innervate the upper limbs. What is the name of this enlarged region?

_____

_____

BEYOND PAL **7.** The brain stem is a made up of which three major structures of the brain?

_____

_____

_____

8. What fluid is contained within the ventricles of the brain, and which structure is responsible for producing this fluid?

_____

_____

9. What hormone is secreted by the pineal gland?

_____

_____

10. The myelencephalon is a developmental region of the brain that becomes which structure of the adult brain?

_____

_____

II. Apply What You Learned

1. In order to perform its function of smoothing and coordinating body movements, the cerebellum needs to integrate what three major pieces of information?

_____

_____

_____

_____

_____

_____

_____

2. If the cerebellum is damaged, how would that affect body movements? Why?

_____

_____

_____

_____

_____

_____

_____

_____

_____

_____

SELF REVIEW | Anatomical Models

Exercise 4.14 **Brain Regions**

GO TO ANATOMICAL MODELS > NERVOUS SYSTEM > CENTRAL NERVOUS SYSTEM > SELF REVIEW > IMAGE 1

- *Mouse over the image to locate and label the structures indicated below. Click on the structures to hear their pronunciations.*

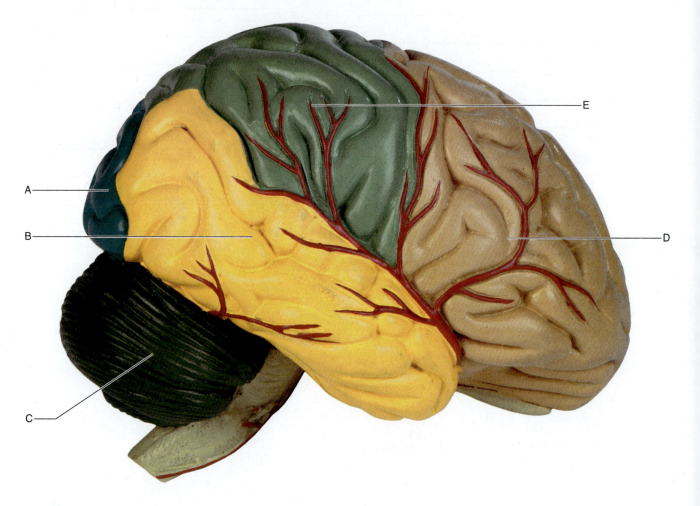

A. _____

B. _____

C. _____

D. _____

E. _____

Exercise 4.15   **Midsagittal Brain**

GO TO ANATOMICAL MODELS > NERVOUS SYSTEM > CENTRAL NERVOUS SYSTEM > SELF REVIEW >
IMAGES 2 AND 3

- *Mouse over the images to locate and label the structures indicated below. Click on the structures to hear their pronunciations.*

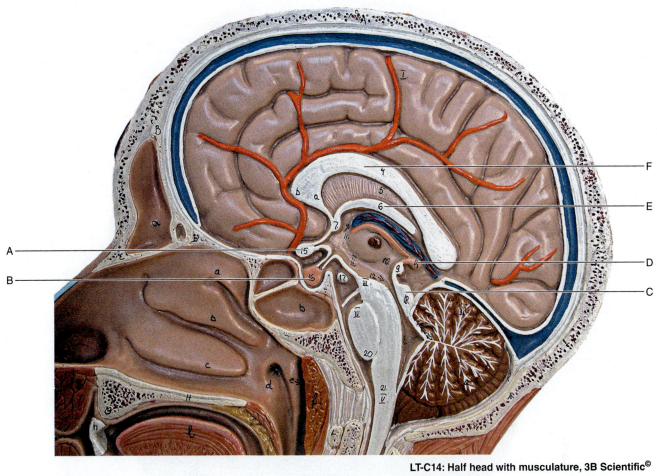

LT-C14: Half head with musculature, 3B Scientific©

A. _____   D. _____

B. _____   E. _____

C. _____   F. _____

Exercise 4.16   **Brain, Inferior View**

GO TO ⟩ ANATOMICAL MODELS > NERVOUS SYSTEM > CENTRAL NERVOUS SYSTEM > SELF REVIEW > IMAGE 4

• *Mouse over the image to locate and label the structures indicated below. Click on the structures to hear their pronunciations.*

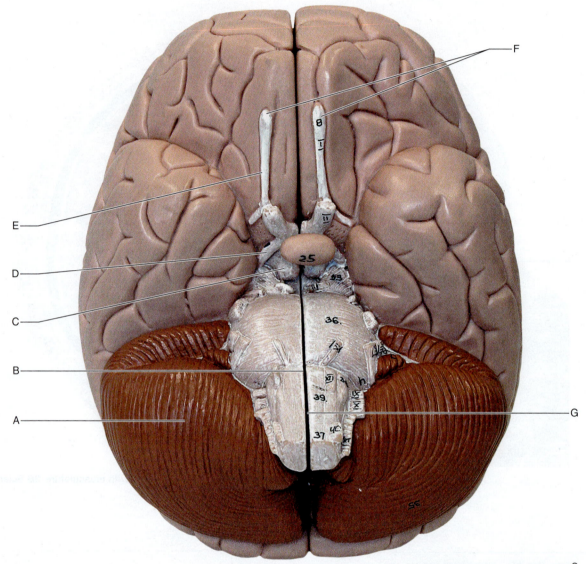

LT-C15: Brain, 2-part, 3B Scientific®

A. _____   E. _____

B. _____   F. _____

C. _____   G. _____

D. _____

Exercise 4.17   **Diencephalon and Brain Stem**

GO TO ANATOMICAL MODELS > NERVOUS SYSTEM > CENTRAL NERVOUS SYSTEM > SELF REVIEW >
IMAGES 7, 8, AND 14

- *Mouse over the images to locate and label the structures indicated below. Click on the structures to hear their pronunciations.*

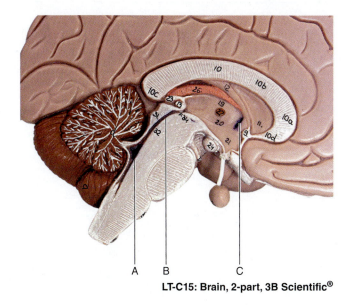

LT-C15: Brain, 2-part, 3B Scientific®

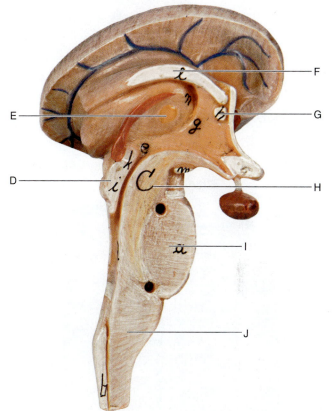

Copyright by SOMSO, 2010, www.somso.com

A. _____   F. _____

B. _____   G. _____

C. _____   H. _____

D. _____   I. _____

E. _____   J. _____

Exercise 4.18 **Spinal Cord**

GO TO〉 ANATOMICAL MODELS > NERVOUS SYSTEM > CENTRAL NERVOUS SYSTEM > SELF REVIEW > IMAGE 16

- *Mouse over the image to locate and label the structures indicated below. Click on the structures to hear their pronunciations.*

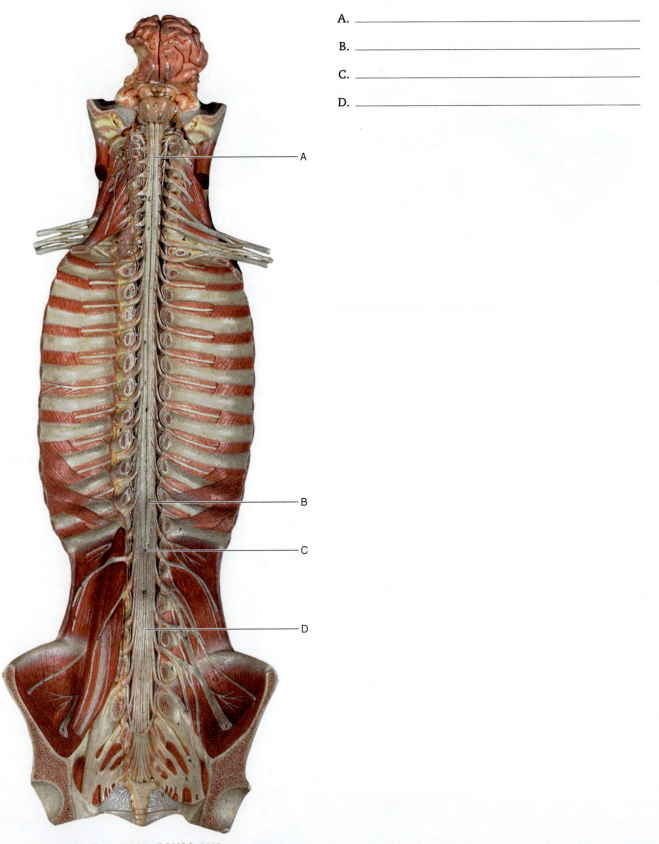

A. _____

B. _____

C. _____

D. _____

Exercise 4.19 Spinal Cord, Cross Section

GO TO ANATOMICAL MODELS > NERVOUS SYSTEM > CENTRAL NERVOUS SYSTEM > SELF REVIEW >
IMAGES 17 AND 18

- *Mouse over the images to locate and label the structures indicated below. Click on the structures to hear their pronunciations.*

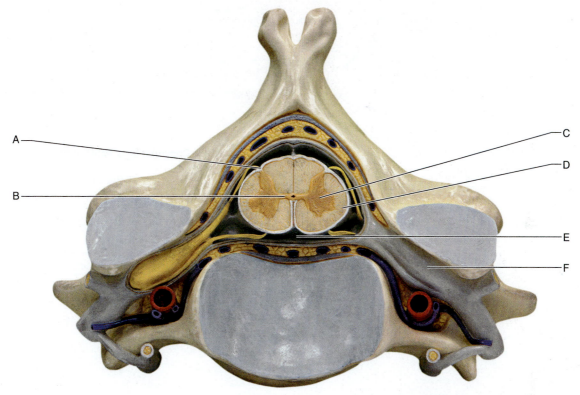

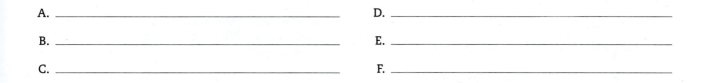

A. _____   D. _____

B. _____   E. _____

C. _____   F. _____

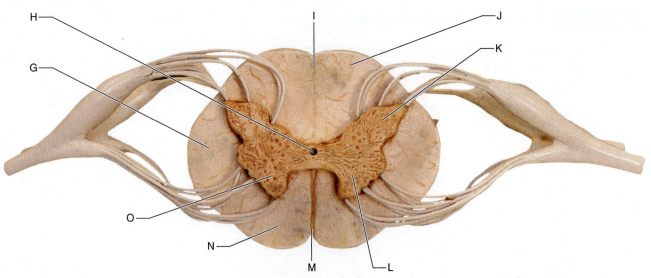

G. _____   L. _____

H. _____   M. _____

I. _____   N. _____

J. _____   O. _____

K. _____

QUIZ Anatomical Models

Answer the following questions using information from PAL 3.1 as well as other course materials including your textbook, lecture, and lab notes.

I. Check Your Understanding

1. Name the five paired cerebral lobes.

_____

_____

_____

_____

_____

2. The midbrain, pons, and medulla oblongata together comprise what region of the brain?

_____

_____

3. The diencephalon contains which ventricle of the brain? What fluid is contained within this ventricle?

_____

_____

4. The corpus callosum is an example of what type of fiber tract?

_____

5. What is the dural septa that separates the cerebrum from the cerebellum?

_____

6. Name the three subregions of the diencephalon.

_____

_____

_____

BEYOND
PAL

7. Which structures of the arachnoid mater are responsible for the resorption of cerebrospinal fluid from the subarachnoid space into the dural sinuses?

_____

_____

8. The telencephalon is a developmental region of the brain that becomes which structure of the adult brain?

_____

9. Which region of the spinal cord contains neuronal cell bodies of somatic motor neurons in the peripheral nervous system?

_____

10. The primary visual cortex is found in which cerebral lobe?

_____

II. Apply What You Learned

After falling off her bike and hitting her helmeted head on the pavement, Jackie is told she may have a concussion and needs to remain in the hospital overnight for observation. Jackie has a headache, but otherwise feels fine and is irritated that she can't go home. As her doctor, you must explain to Jackie why a concussion can be dangerous. How would you answer each of Jackie's arguments as to why she should be able to leave the hospital?

1. "I was wearing a helmet, and I don't have any cuts on my head, so there can't possibly be anything major wrong with me."

_____

_____

_____

_____

_____

_____

_____

_____

2. "I only have a headache. If I had a concussion wouldn't there be other problems?"

_____

_____

_____

_____

_____

_____

3. "If I had a concussion, wouldn't the damage be immediately apparent?"

_____

_____

_____

_____

_____

THE PERIPHERAL NERVOUS SYSTEM

SELF REVIEW | Human Cadaver

Exercise 4.20 **Cranial Nerves**

 HUMAN CADAVER > NERVOUS SYSTEM > PERIPHERAL NERVOUS SYSTEM > SELF REVIEW > IMAGE 1

- *Mouse over the image to locate and label the structures indicated below. Click on the structures to hear their pronunciations.*

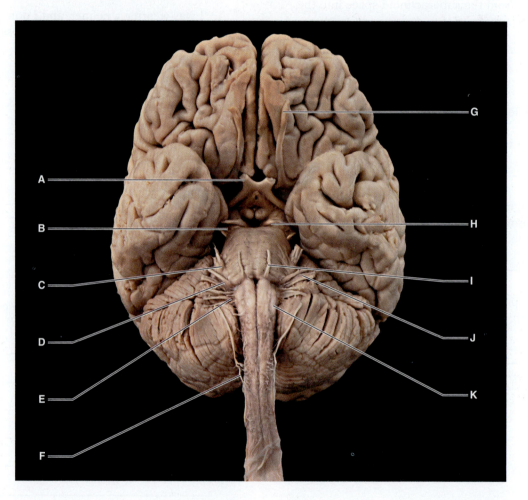

A. _____   G. _____

B. _____   H. _____

C. _____   I. _____

D. _____   J. _____

E. _____   K. _____

F. _____

Exercise 4.21   **Spinal Nerves**

GO TO > HUMAN CADAVER > NERVOUS SYSTEM > PERIPHERAL NERVOUS SYSTEM > SELF REVIEW > IMAGE 7

- *Mouse over the image to locate and label the structures associated with the spinal nerves. Click on the structures to hear their pronunciations.*

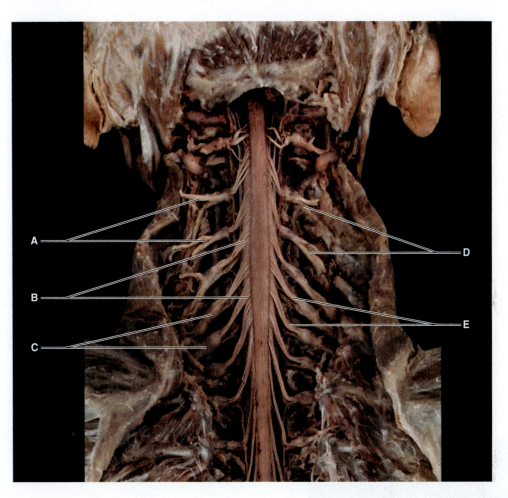

A. _____   D. _____

B. _____   E. _____

C. _____

Exercise 4.22   **Brachial Plexus**

GO TO › HUMAN CADAVER > NERVOUS SYSTEM > PERIPHERAL NERVOUS SYSTEM > SELF REVIEW > IMAGES 9 AND 10

- *Mouse over the images to locate and label the structures indicated below. Click on the structures to hear their pronunciations.*

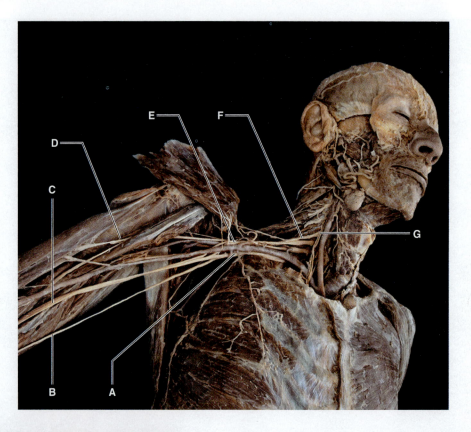

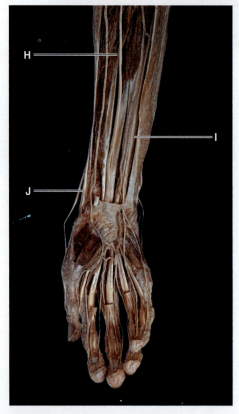

A. _____

B. _____

C. _____

D. _____

E. _____

F. _____

G. _____

H. _____

I. _____

J. _____

Exercise 4.23 Nerves of Lumbar and Sacral Plexuses

GO TO HUMAN CADAVER > NERVOUS SYSTEM > PERIPHERAL NERVOUS SYSTEM > SELF REVIEW > IMAGES 11 AND 12

> • *Mouse over the images to locate and label the structures indicated below. Click on the structures to hear their pronunciations.*

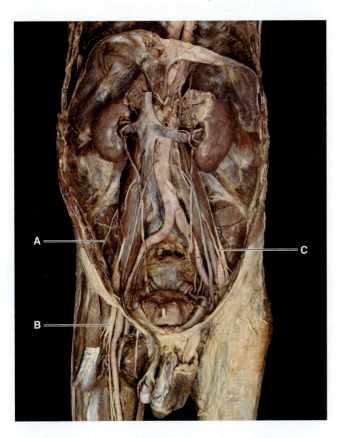

A. _____   D. _____

B. _____   E. _____

C. _____   F. _____

QUIZ Human Cadaver

Answer the following questions using information from PAL 3.1 as well as other course materials including your textbook, lecture, and lab notes.

I. Check Your Understanding

1. How many pairs of cranial nerves are there?

_____

2. Which cranial nerve carries motor information to the superior oblique muscle of the eye?

_____

3. The phrenic nerve innervates the diaphragm. From which nerve plexus does it originate?

_____

4. The musculocutaneous nerve is a branch of which nerve plexus?

_____

BEYOND PAL **5.** Which cranial nerve is the only one to have innervation function outside the head and neck?

_____

6. What are the three divisions of the trigeminal nerve?

_____

_____

7. Identify each of the following as a mechanoreceptor, thermoreceptor, photoreceptor, chemoreceptor, or nociceptor.

Meissner's
corpuscle   _____

Free nerve ending
pain receptor   _____

Free nerve ending
temperature receptor   _____

Spiral organ
of Corti receptors   _____

Proprioceptors   _____

Rods and cones
of retina   _____

8. The cell bodies of somatic sensory neurons are found within which structures?

_____

II. Apply What You Learned

Polio (poliomyelitis) is a disease that is caused by a virus that attacks the nervous system. It can result in temporary paralysis or permanent paralysis of skeletal muscles, including the diaphragm.

1. What structures of the nervous system is the virus preferentially attacking?

_____

_____

2. Why do some muscles have the ability to regain function?

_____

_____

_____

_____

_____

_____

_____

3. Which nerve does the virus have to affect to cause paralysis of the diaphragm?

_____

_____

SELF REVIEW | Anatomical Models

Exercise 4.24 Cranial Nerves

GO TO ANATOMICAL MODELS > NERVOUS SYSTEM > PERIPHERAL NERVOUS SYSTEM > SELF REVIEW > IMAGE 2

- *Mouse over the image to locate and label the structures indicated below. Click on the structures to hear their pronunciations.*

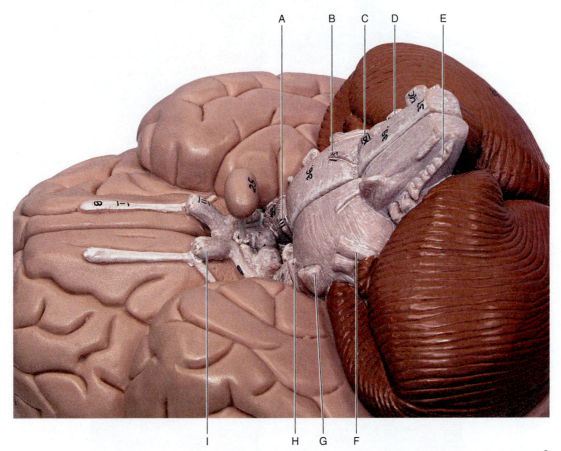

LT-C15: Brain, 2-part, 3B Scientific®

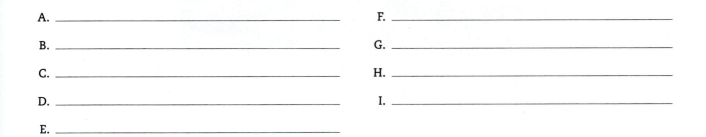

A. _____   F. _____

B. _____   G. _____

C. _____   H. _____

D. _____   I. _____

E. _____

Exercise 4.25 **Spinal Nerves**

GO TO ANATOMICAL MODELS > NERVOUS SYSTEM > PERIPHERAL NERVOUS SYSTEM > SELF REVIEW > IMAGE 5

• *Mouse over the image to locate and label the structures indicated below. Click on the structures to hear their pronunciations.*

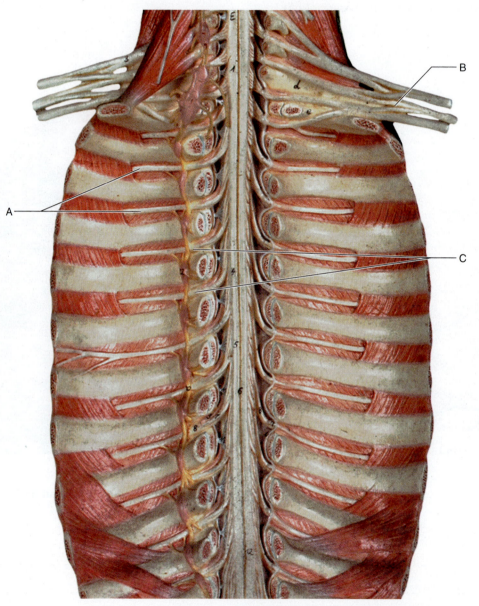

Copyright by SOMSO, 2010, www.somso.com

A. _____   C. _____

B. _____

Exercise 4.26   **Brachial Plexus**

GO TO ANATOMICAL MODELS > NERVOUS SYSTEM > PERIPHERAL NERVOUS SYSTEM > SELF REVIEW >
IMAGES 7 AND 8

- *Mouse over the images to locate and label the structures indicated below. Click on the structures to hear their pronunciations.*

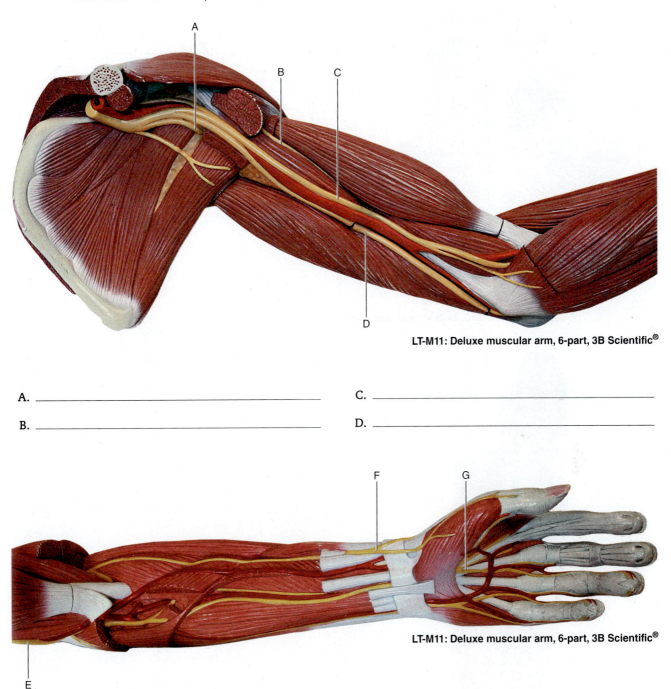

LT-M11: Deluxe muscular arm, 6-part, 3B Scientific®

A. _____

B. _____

C. _____

D. _____

LT-M11: Deluxe muscular arm, 6-part, 3B Scientific®

E. _____

F. _____

G. _____

Exercise 4.27 **Lumbar and Sacral Plexuses**

GO TO ANATOMICAL MODELS > NERVOUS SYSTEM > PERIPHERAL NERVOUS SYSTEM > SELF REVIEW > IMAGES 6 AND 9

- *Mouse over the images to locate and label the structures indicated below. Click on the structures to hear their pronunciations.*

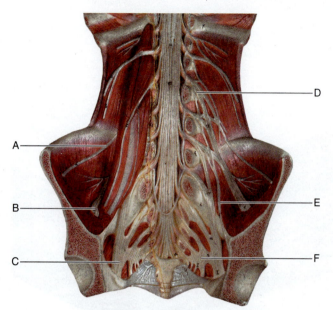

Copyright by SOMSO, 2010, www.somso.com

A. _____

B. _____

C. _____

D. _____

E. _____

F. _____

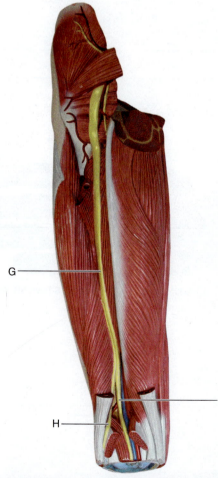

G. _____

H. _____

I. _____

LT-M21: Muscular leg, 7-part, 3B Scientific®

QUIZ Anatomical Models

Answer the following questions using information from PAL 3.1 as well as other course materials including your textbook, lecture, and lab notes.

I. Check Your Understanding

1. How many pairs of spinal nerves are there?

_____

2. Which nerve plexus is responsible for innervating muscles on the anterior aspect of the thigh?

_____

3. Which spinal nerves form the brachial plexus?

_____

_____

4. In the schematic below, correctly label the spinal cord, spinal nerve, dorsal ramus, ventral ramus, dorsal root ganglion, dorsal root, and ventral root.

5. The radial nerve is a branch of which nerve plexus?

_____

BEYOND PAL **6.** Which cranial nerves transmit only sensory information?

_____

_____

7. Which division of the trigeminal nerve contains motor fibers that innervate chewing muscles?

_____

8. Which cranial nerve transmits hearing and equilibrium information from the inner ear?

_____

II. Apply What You Learned

1. Spinal cord damage that occurs at level T3 will impact the function of which plexuses?

_____

_____

2. How would spinal cord damage affect sensory input below the T3 level?

_____

_____

_____

_____

3. How would spinal cord damage affect somatic motor output to regions below the T3 level?

_____

_____

_____

_____

THE AUTONOMIC NERVOUS SYSTEM

SELF REVIEW | Human Cadaver

Exercise 4.28 **Brain Structures**

GO TO > HUMAN CADAVER > NERVOUS SYSTEM > AUTONOMIC NERVOUS SYSTEM > SELF REVIEW > IMAGE 1

- *Mouse over the image to locate and label the structures indicated below. Click on the structures to hear their pronunciations.*

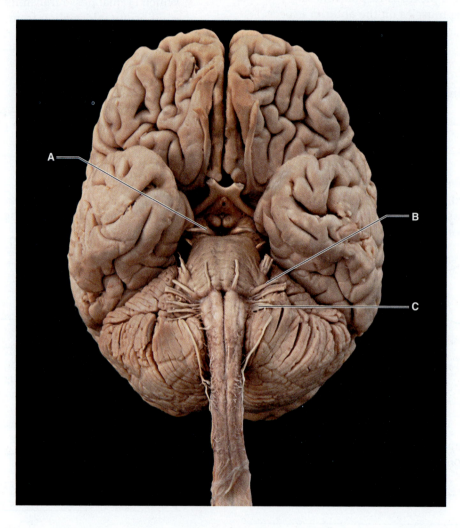

A. _____   C. _____

B. _____

Exercise 4.29   **Autonomic Nervous System Structures**

GO TO › HUMAN CADAVER > NERVOUS SYSTEM > AUTONOMIC NERVOUS SYSTEM > SELF REVIEW > IMAGE 4

- *Mouse over the image and locate and label the structures indicated below. Click on the structures to hear their pronunciations.*

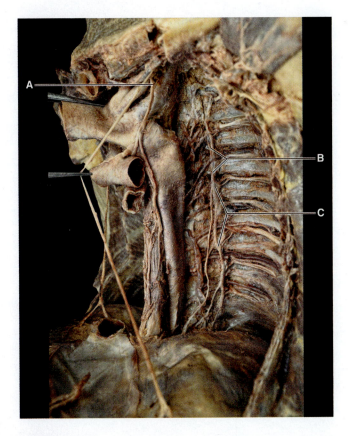

A. _____

B. _____

C. _____

QUIZ) Human Cadaver

Answer the following questions using information from PAL 3.1 as well as other course materials including your textbook, lecture, and lab notes.

I. Check Your Understanding

BEYOND PAL **1.** In which division of the peripheral nervous system is the autonomic nervous system?

_____

2. Which division of the autonomic nervous system is known for overseeing "fight, flight, or fright" responses?

_____

3. Which neurotransmitter is released by the preganglionic fibers of the autonomic nervous system?

_____

4. The adrenal gland is part of which division of the autonomic nervous system? Which hormones are released by the modified postganglionic neurons of the adrenal medulla?

_____

_____

_____

II. Apply What You Learned

1. After suffering spinal cord damage to region L1 of his spinal cord, Simon is told he may or may not regain complete function, but that the parasympathetic division of his autonomic nervous system was not affected. Simon has taken a course in human anatomy and knows this is not necessarily true. Why?

_____

_____

_____

_____

_____

_____

_____

_____

_____

2. Indicate which division of the autonomic nervous system elicits the responses listed below.

RESPONSE	SYMPATHETIC DIVISION	PARASYMPATHETIC DIVISION
↑ bronchiolar constriction		
↑ heart rate		
↑ metabolism		
↑ gastrointestinal motility		
↑ pupillary diameter		
↑ gallbladder contraction		
↑ blood glucose		
↑ salivation		
↑ coagulation		
↑ contraction of bladder wall		

SELF REVIEW | Anatomical Models

Exercise 4.30 Autonomic Nervous System Structures

GO TO ANATOMICAL MODELS > NERVOUS SYSTEM > AUTONOMIC NERVOUS SYSTEM > SELF REVIEW > IMAGES 2 AND 3

- *Mouse over the images to locate and label the structures indicated below. Click on the structures to hear their pronunciations.*

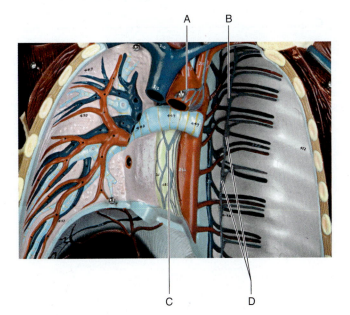

A. _____

B. _____

C. _____

D. _____

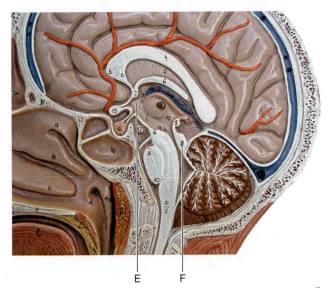

LT-C14: Half head with musculature, 3B Scientific©

E. _____

F. _____

Exercise 4.31   **Autonomic Nervous System, Cranial Nerves**

GO TO ANATOMICAL MODELS > NERVOUS SYSTEM > AUTONOMIC NERVOUS SYSTEM > SELF REVIEW > IMAGE 4

- *Mouse over the image to locate and label the structures indicated below. Click on the structures to hear their pronunciations.*

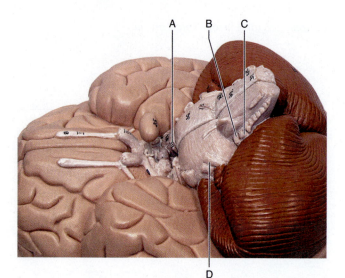

A. _____

B. _____

C. _____

D. _____

LT-C15: Brain, 2-part, 3B Scientific®

Exercise 4.32   **Adrenal Gland**

GO TO ANATOMICAL MODELS > NERVOUS SYSTEM > AUTONOMIC NERVOUS SYSTEM > SELF REVIEW > IMAGE 6

- *Mouse over the image to locate and label the structure indicated below. Click on the structure to hear its pronunciation.*

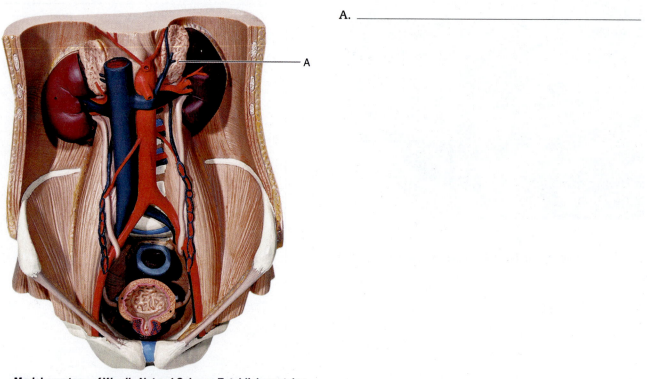

A. _____

Model courtesy of Ward's Natural Science Establishment, Inc.,
www.wardsci.com

QUIZ Anatomical Models

Answer the following questions using information from PAL 3.1 as well as other course materials including your textbook, lecture, and lab notes.

I. Check Your Understanding

BEYOND PAL

1. Is the autonomic nervous system a motor or sensory pathway? Is it visceral or somatic?

_____

_____

2. The postganglionic fibers of which division of the autonomic nervous system typically release norepinephrine?

_____

3. Which division(s) of the autonomic nervous system has preganglionic fibers that originate in cranial nerves?

_____

_____

4. Which division(s) of the autonomic nervous system typically have preganglionic fibers that are longer than postganglionic fibers?

_____

_____

5. Identify one peripheral structure that is served by the sympathetic division but has NO innervations from the parasympathetic division.

_____

II. Apply What You Learned

1. Continual stress results in over-activation of which division of the autonomic nervous system?

_____

2. What portion of a blood vessel wall is under sympathetic control? How does the blood vessel respond to sympathetic innervations?

_____

_____

_____

_____

3. Continual stress can result in a dangerous medical condition called hypertension because of the way it impacts blood vessel function. What are the long-term impacts of hypertension on the heart and arteries?

_____

_____

_____

_____

_____

_____

THE SPECIAL SENSES

SELF REVIEW | Human Cadaver

Exercise 4.33   **Structures of Taste and Hearing**

GO TO › HUMAN CADAVER > NERVOUS SYSTEM > SPECIAL SENSES > SELF REVIEW > IMAGES 3 AND 14

- *Mouse over the images to locate and label the structures indicated below. Click on the structures to hear their pronunciations.*

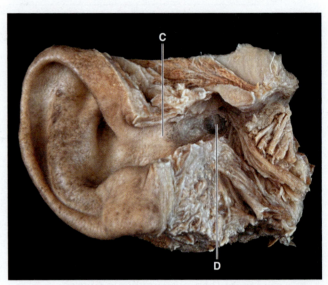

A. _____   C. _____

B. _____   D. _____

Exercise 4.34 Optic Structures

GO TO ⟩ HUMAN CADAVER > NERVOUS SYSTEM > SPECIAL SENSES > SELF REVIEW > IMAGES 6 AND 8

- *Mouse over the images to locate and label the structures indicated below. Click on the structures to hear their pronunciations.*

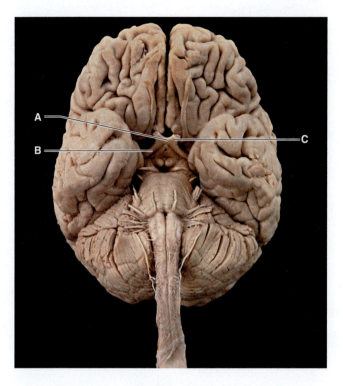

A. _____

B. _____

C. _____

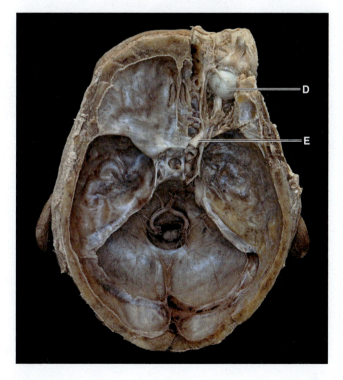

D. _____

E. _____

Exercise 4.35 **External Eye**

GO TO HUMAN CADAVER > NERVOUS SYSTEM > SPECIAL SENSES > SELF REVIEW > IMAGE 11

- *Mouse over the image to locate and label the structures indicated below. Click on the structures to hear their pronunciations.*

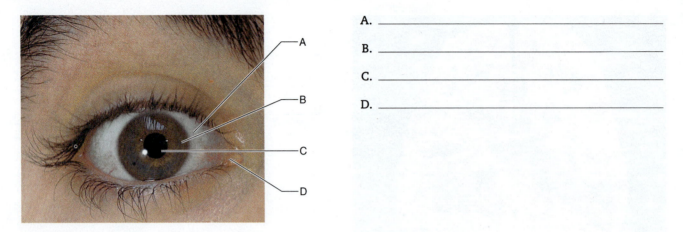

A. _____

B. _____

C. _____

D. _____

Exercise 4.36 **Cow Eye**

GO TO CAT > NERVOUS SYSTEM > SELF REVIEW > IMAGES 17 AND 18

- *Mouse over the images to locate and label the structures indicated below. Click on the structures to hear their pronunciations.*

A. _____

B. _____

C. _____

D. _____

E. _____

F. _____

G. _____

H. _____

I. _____

J. _____

K. _____

QUIZ ⟩ Human Cadaver

Answer the following questions using information from PAL 3.1 as well as other course materials including your textbook, lecture, and lab notes.

I. Check Your Understanding

1. Which papillae of the tongue contain taste buds?

   _____

2. What are the five special senses?

   _____

   _____

   _____

   _____

   _____

3. Which structure of the eye contains photoreceptors?

   _____

4. Which structure separates the external ear from the middle ear, and functions to transmit sound waves to the auditory ossicles?

   _____

5. Which layer of the eye is a continuation of the dura mater?

   _____

BEYOND PAL 6. Olfactory receptor cells are what type of neuron (unipolar, bipolar, or multipolar)?

   _____

II. Apply What You Learned

1. Middle ear infections (otitis media) are a common ailment in small children. Which structure in the developing ear allows viruses and bacteria to easily access the middle ear? How does this developing structure differ anatomically from the adult structure?

   _____

   _____

   _____

   _____

   _____

   _____

2. Why is inserting a tube through the tympanic membrane helpful in cases of chronic ear infections?

   _____

   _____

   _____

   _____

3. If the medical term for a middle ear infection is "otitis media," what is the proper term for an infection in the external auditory canal?

   _____

SELF REVIEW | Anatomical Models

Exercise 4.37 External Eye and Extrinsic Eye Muscles

GO TO ANATOMICAL MODELS > NERVOUS SYSTEM > SPECIAL SENSES > SELF REVIEW > IMAGES 1 AND 7

- *Mouse over the images to locate and label the structures indicated below. Click on the structures to hear their pronunciations.*

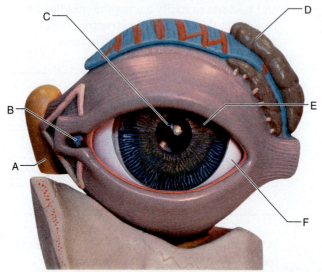

A. _____

B. _____

C. _____

D. _____

E. _____

F. _____

LT-F12: Giant eye with eyelid and lachrymal system, 8-part, 3B Scientific®

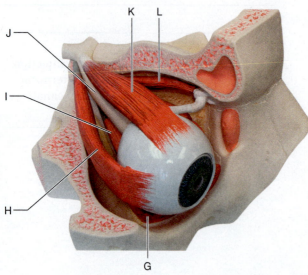

G. _____

H. _____

I. _____

J. _____

K. _____

L. _____

LT-F13: Classic eye in orbit 7-part, 3B Scientific®

Exercise 4.38   Internal Structures of the Eye

GO TO ⟩ ANATOMICAL MODELS > NERVOUS SYSTEM > SPECIAL SENSES > SELF REVIEW > IMAGES 3 AND 4

- *Mouse over the images to locate and label the structures indicated below. Click on the structures to hear their pronunciations.*

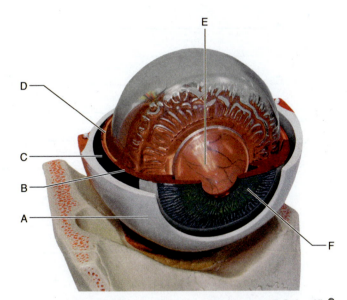

LT-F12: Giant eye with eyelid and lachrymal system, 8-part, 3B Scientific®

A. _____

B. _____

C. _____

D. _____

E. _____

F. _____

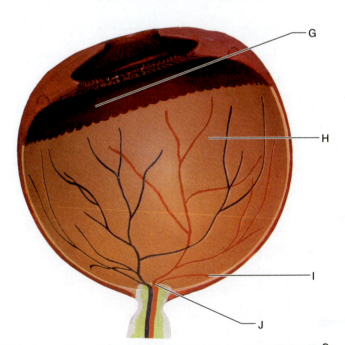

LT-F12: Giant eye with eyelid and lachrymal system, 8-part, 3B Scientific®

G. _____

H. _____

I. _____

J. _____

Exercise 4.39   **Structures of the Ear**

GO TO ANATOMICAL MODELS > NERVOUS SYSTEM > SPECIAL SENSES > SELF REVIEW > IMAGES 9 AND 10

- *Mouse over the images to locate and label the structures indicated below. Click on the structures to hear their pronunciations.*

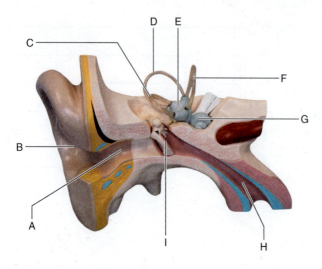

A. _____

B. _____

C. _____

D. _____

E. _____

F. _____

G. _____

H. _____

I. _____

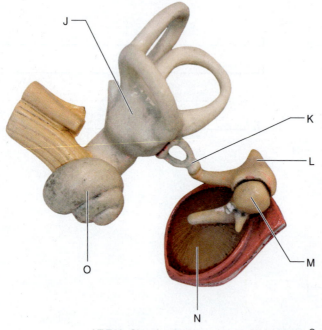

J. _____

K. _____

L. _____

M. _____

N. _____

O. _____

LT-E10: Classic giant ear, 4-part, 3B Scientific®

Exercise 4.40   **Cochlea and Spiral Organ of Corti**

GO TO ⟩ ANATOMICAL MODELS > NERVOUS SYSTEM > SPECIAL SENSES > SELF REVIEW > IMAGES 12 AND 13

- *Mouse over the images to locate and label the structures indicated below. Click on the structures to hear their pronunciations.*

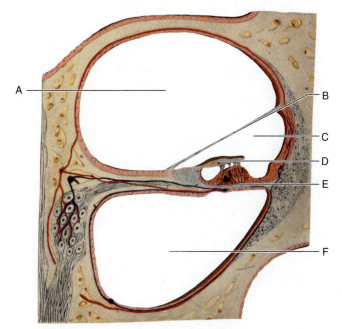

A. _____

B. _____

C. _____

D. _____

E. _____

F. _____

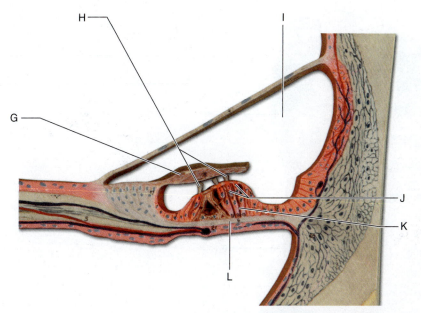

G. _____   J. _____

H. _____   K. _____

I. _____   L. _____

Exercise 4.41 **Structures of Smell and Taste**

GO TO ANATOMICAL MODELS > NERVOUS SYSTEM > SPECIAL SENSES > SELF REVIEW > IMAGES 14 AND 15

- *Mouse over the images to locate and label the structures indicated below. Click on the structures to hear their pronunciations.*

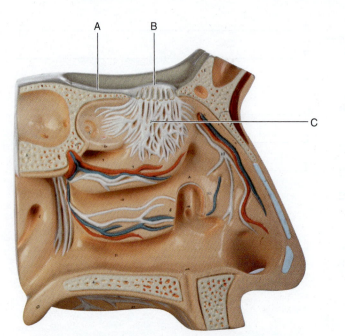

Model courtesy of Denoyer-Geppert, www.denoyer.com

A. _____

B. _____

C. _____

D. _____

E. _____

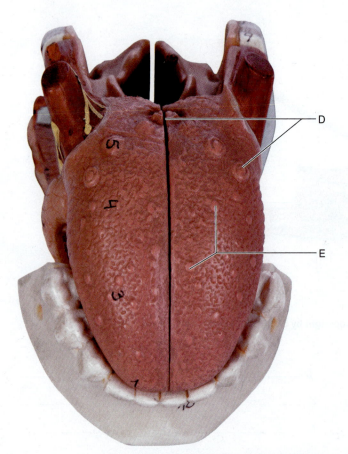

QUIZ Anatomical Models

Answer the following questions using information from PAL 3.1 as well as other course materials including your textbook, lecture, and lab notes.

I. Check Your Understanding

1. Olfactory receptor cells synapse within glomeruli of which central nervous system structure?

   _____

2. Which structure of the cochlear duct contains the receptor cells for hearing?

   _____

3. Which cranial nerve transmits sensory impulses from the eye to the brain?

   _____

BEYOND PAL
4. Which cells of a taste bud are chemoreceptors?

   _____

5. Which of the special senses depend on mechanoreceptors?

   _____

   _____

6. Which membranous structures of the inner ear house the cristae ampullares?

   _____

   _____

II. Apply What You Learned

1. Anosmia is a loss of smell. One of the primary causes of anosmia is a head injury. If the cribriform plate is damaged by a blow to the head, how could this result in anosmia?

   _____

   _____

   _____

   _____

   _____

2. For each special sense listed below, indicate which cranial nerves transmit the sensory input to the brain. Also indicate which lobe of the cerebral cortex processes the input.

SPECIAL SENSE	CRANIAL NERVES	LOBE
Vision		
Olfaction		
Taste		
Hearing		
Equilibrium		

TISSUES OF THE NERVOUS SYSTEM

SELF REVIEW | Histology

Exercise 4.42 **Multipolar Neuron Smear 400x**

GO TO ⟩ HISTOLOGY > NERVOUS TISSUE > SELF REVIEW > IMAGE 2

> • *Mouse over the image to locate and label the structures indicated below. Click on the structures to hear their pronunciations.*

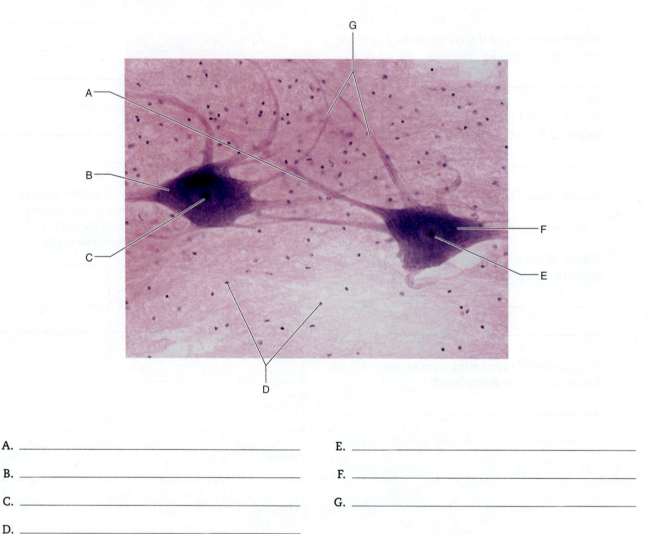

A. _____   E. _____

B. _____   F. _____

C. _____   G. _____

D. _____

Exercise 4.43   **Nerve, Longitudinal Section 100x and 400x**

GO TO ⟩ HISTOLOGY > NERVOUS TISSUE > SELF REVIEW > IMAGES 7 AND 8

- *Mouse over the images to locate and label the structures indicated below. Click on the structures to hear their pronunciations.*

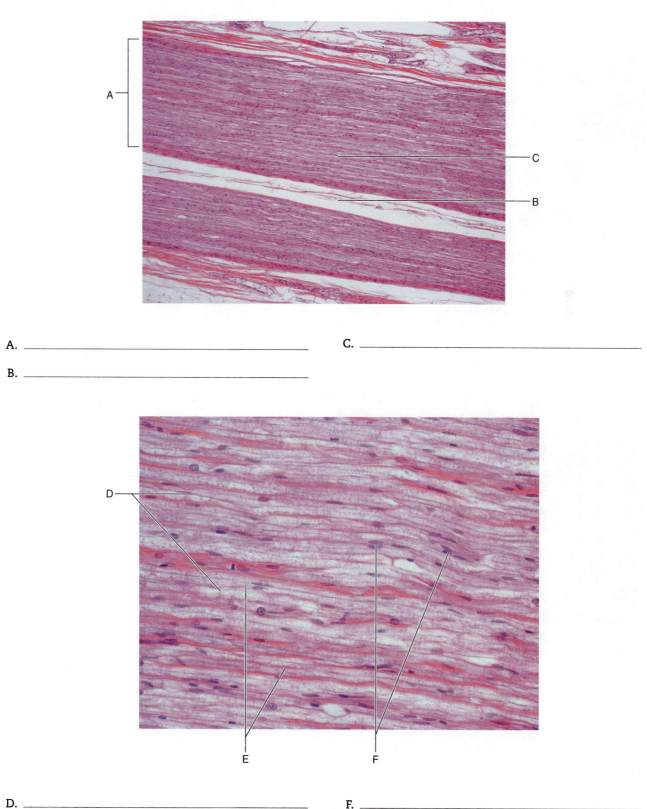

A. _____   C. _____

B. _____

D. _____   F. _____

E. _____

Exercise 4.44 **Nerve, Cross Section 100x and 400x**

GO TO HISTOLOGY > NERVOUS TISSUE > SELF REVIEW > IMAGES 9 AND 10

- *Mouse over the images to locate and label the structures indicated below. Click on the structures to hear their pronunciations.*

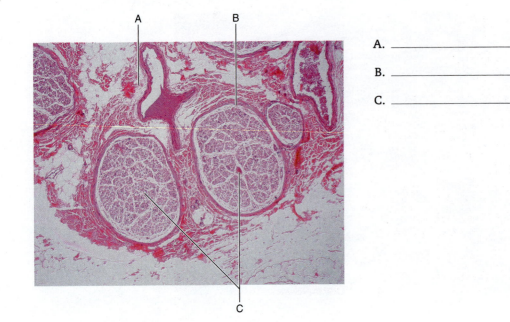

A. _____

B. _____

C. _____

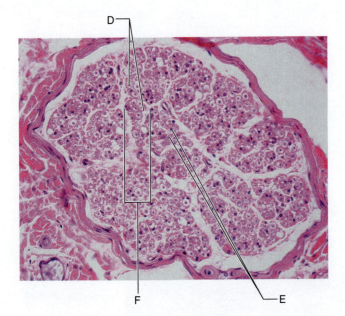

D. _____

E. _____

F. _____

Exercise 4.45   **Cerebellum, Silver Stain, Sagittal Section 40x and 200x**

GO TO   HISTOLOGY > NERVOUS TISSUE > SELF REVIEW > IMAGES 12 AND 14

> • *Mouse over the images to locate and label the structures indicated below. Click on the structures to hear their pronunciations.*

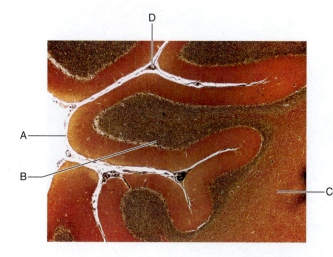

A. _____

B. _____

C. _____

D. _____

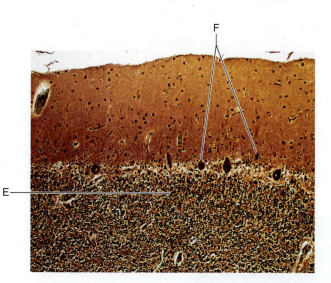

E. _____

F. _____

Exercise 4.46 **Spinal Cord, Cross Section, Cervical and Thoracic Region 20x**

GO TO ⟩ HISTOLOGY > NERVOUS TISSUE > SELF REVIEW > IMAGES 15 AND 18

• *Mouse over the images to locate and label the structures indicated below. Click on the structures to hear their pronunciations.*

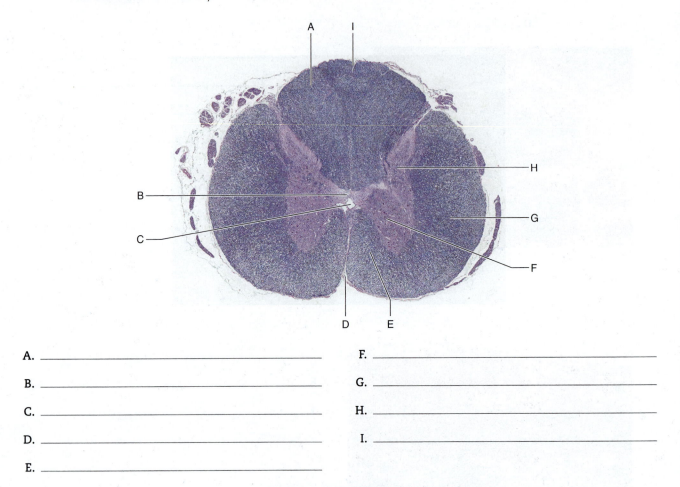

A. _____   F. _____

B. _____   G. _____

C. _____   H. _____

D. _____   I. _____

E. _____

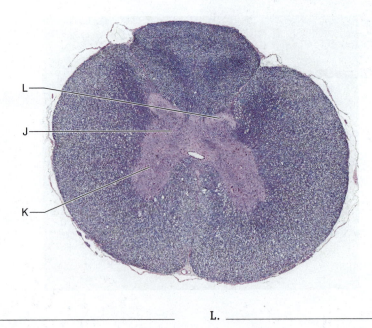

J. _____   L. _____

K. _____

Exercise 4.47 **Dorsal Root Ganglion, Cross Section 400x**

GO TO HISTOLOGY > NERVOUS TISSUE > SELF REVIEW > IMAGE 24

- *Mouse over the image to locate and label the structures indicated below. Click on the structures to hear their pronunciations.*

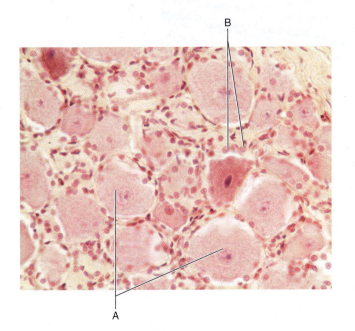

A. _____

B. _____

Exercise 4.48 **Retina, Cross Section 100x, 200x and Optic Nerve, Cross Section 40x**

GO TO HISTOLOGY > SPECIAL SENSES > SELF REVIEW > IMAGES 1, 2, AND 5

- *Mouse over the images to locate and label the structures indicated below. Click on the structures to hear their pronunciations.*

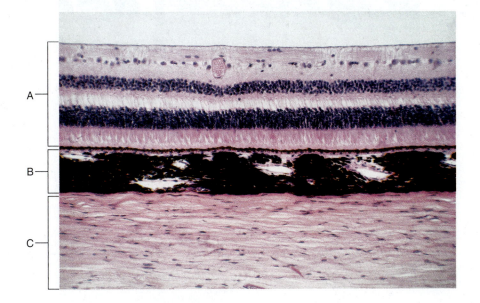

A. _____   C. _____

B. _____

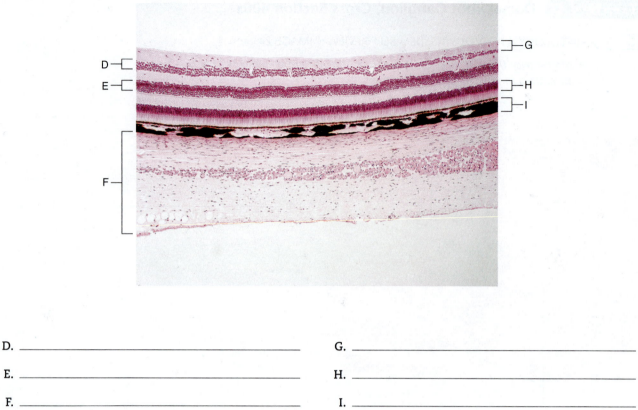

D. _____   G. _____

E. _____   H. _____

F. _____   I. _____

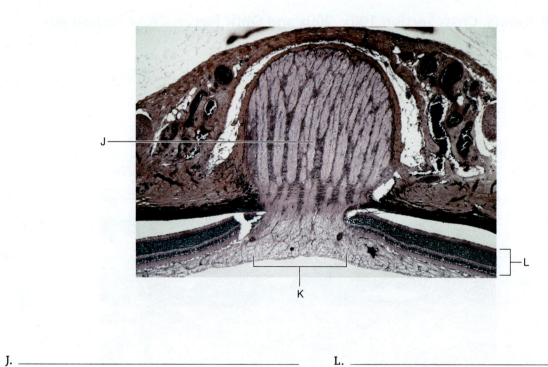

J. _____   L. _____

K. _____

Exercise 4.49   **Cochlea, Cross Section 40x, 100x, and 400x**

GO TO HISTOLOGY > SPECIAL SENSES > SELF REVIEW > IMAGES 6, 7, AND 9

• *Mouse over the images to locate and label the structures indicated below. Click on the structures to hear their pronunciations.*

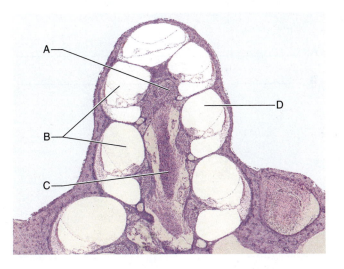

A. _____

B. _____

C. _____

D. _____

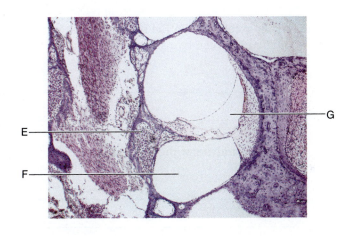

E. _____

F. _____

G. _____

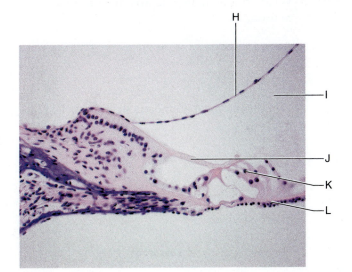

H. _____

I. _____

J. _____

K. _____

L. _____

Exercise 4.50 **Taste Bud, Cross Section 200x**

GO TO HISTOLOGY > SPECIAL SENSES > SELF REVIEW > IMAGE 12

- *Mouse over the image to locate and label the structures indicated below. Click on the structures to hear their pronunciations.*

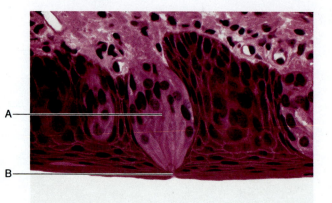

A. _____

B. _____

QUIZ Histology

Answer the following questions using information from PAL 3.1 as well as other course materials including your textbook, lecture, and lab notes.

I. Check Your Understanding

1. What is the name of the tapered end of a neuronal cell body?

_____

2. What is the term used for gaps in the myelin sheath?

_____

3. What is the term used for sensory neurons that transmit nerve impulses from sensory receptors toward the central nervous system?

_____

_____

4. The eyeball has three layers. Name these layers.

_____

_____

_____

5. The hair cells within the scala media are situated between two structures. Name these structures.

_____

_____

6. Which chambers of the cochlea contain perilymph?

_____

_____

BEYOND PAL **7.** What are the main receptive regions of a neuron?

_____

_____

8. What is the most abundant structural class of neurons?

_____

9. What cells of the retina respond to light? Which type of light-sensitive cell is important for color vision?

_____

_____

10. What is the optic disc? Why is it also known as the blind spot?

_____

_____

_____

_____

11. Which division of cranial nerve VIII transmits sensory information from the spiral organ of Corti to the brain?

_____

12. List the three functions of the myelin sheath.

_____

_____

_____

_____

II. Apply What You Learned

1. Describe how multiple sclerosis demyelinates axons. What does the demyelination do to axon potential propagation? What are some of the common symptoms of multiple sclerosis?

_____

_____

_____

_____

_____

_____

_____

_____

_____

_____

_____

2. Aunt Sue is a robust, active, and healthy 81-year-old woman. However, she visits her clinician because she is having some problems with her vision. She sees flashes of light, floating debris, and shadows in her field of vision. Her vision is also a bit blurry. Her clinician's diagnosis is retinal detachment. Given Aunt Sue's age, what is the most probable cause of the detachment?

_____

_____

_____

_____

_____

_____

_____

_____

_____

_____

_____

_____

LAB PRACTICAL   Nervous System

1. Identify the structure.

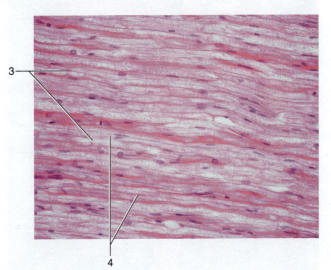

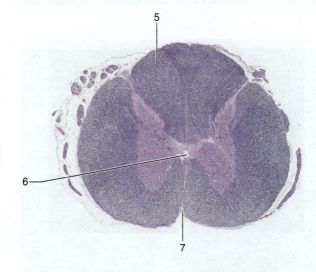

LAB PRACTICAL *continues*

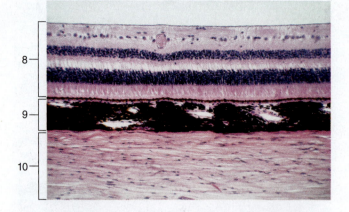

8. Identify the layer.

_____

9. Identify the layer.

_____

10. Identify the layer.

_____

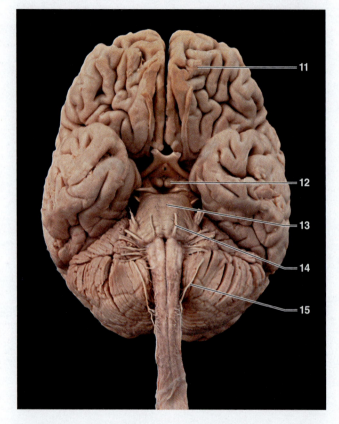

11. Identify the lobe.

_____

12. Identify the structure.

_____

13. Identify the structure.

_____

14. Identify the nerve by name.

_____

15. Identify the nerve by number.

_____

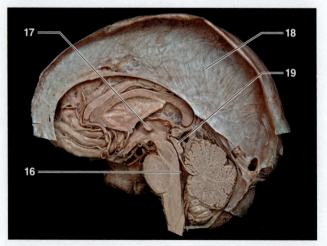

16. Identify the space.

_____

17. Identify the structure.

_____

18. Identify the structure.

_____

19. Identify the structure.

_____

20. Identify the structure.

_____

21. Identify the structure.

_____

22. Identify the structure.

_____

23. Identify the structure.

_____

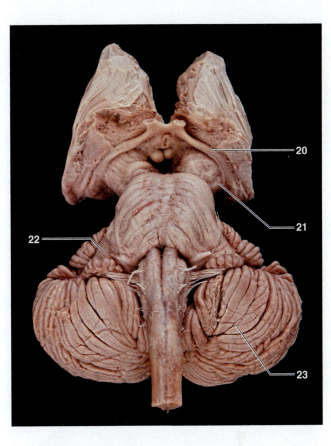

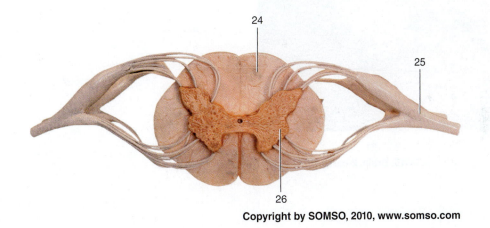

24. Identify the structure.

_____

25. Identify the structure.

_____

26. Identify the structure.

_____

LAB PRACTICAL _continues_

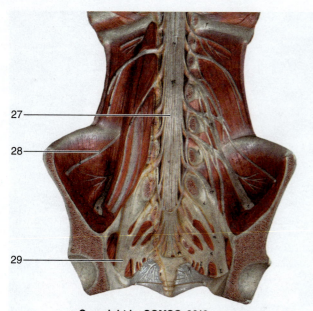

Copyright by SOMSO, 2010, www.somso.com

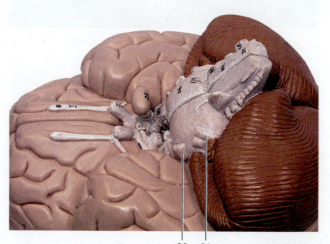

30   31
LT-C15: Brain, 2-part, 3B Scientific®

27. Identify the structure.

_____

28. Identify the nerve.

_____

29. Identify the nerve.

_____

30. Identify the nerve.

_____

31. Identify the nerve.

_____

32. Identify the nerve.

_____

33. Identify the nerve.

_____

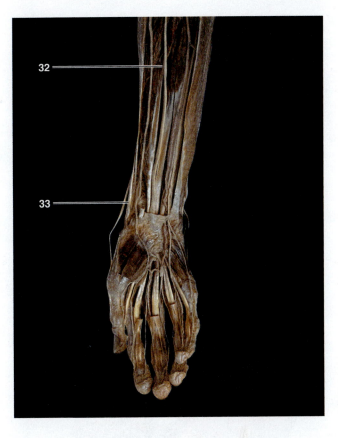

34. Identify the structure.

_____

35. Identify the structure.

_____

36. Identify the structure.

_____

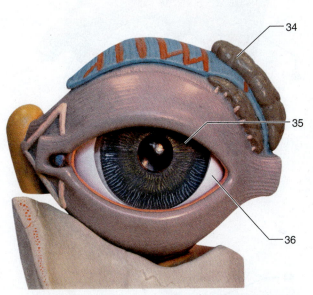

LT-F12: Giant eye with eyelid and lachrymal system, 8-part, 3B Scientific®

LAB PRACTICAL *continues*

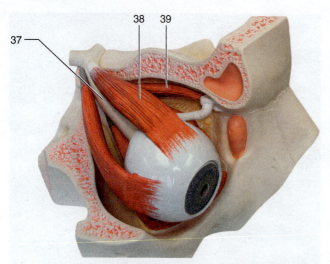

LT-F13: Classic eye in orbit, 7-part, 3B Scientific®

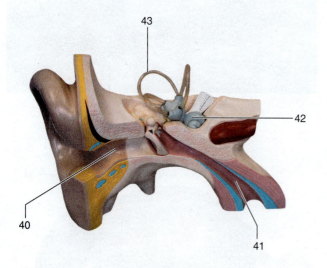

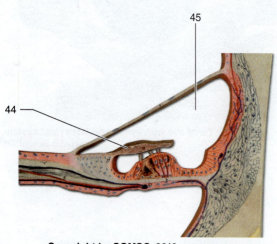

37. Identify the nerve.

_____

38. Identify the structure.

_____

39. Identify the structure.

_____

40. Identify the structure.

_____

41. Identify the structure.

_____

42. Identify the structure.

_____

43. Identify the structure.

_____

44. Identify the structure.

_____

45. Identify the structure.

_____

46. Identify the structure.

_____

47. Identify the structure.

_____

48. Identify the structure.

_____

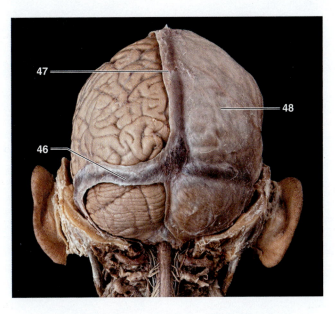

49. Identify the structure.

_____

50. Identify the structure.

_____

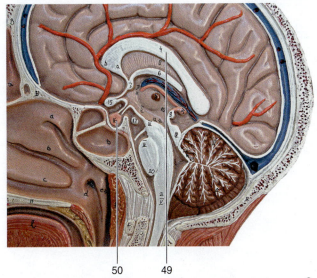

LT-C14: Half head with musculature, 3B Scientific©

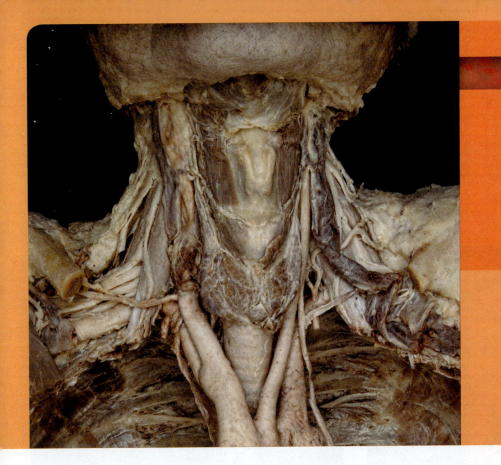

The Endocrine System

STUDENT OBJECTIVES

1. Identify and name the following endocrine structures of the body:
   - Hypothalamus, pituitary gland, and pineal gland
   - Thyroid gland and parathyroid glands
   - Kidneys and adrenal glands
   - Testes
   - Ovaries

2. List the hormones produced and secreted by the endocrine glands.

3. Identify the major functions of hormones produced by the endocrine glands.

4. Describe the relationship between the hypothalamus and the pituitary gland.

GROSS ANATOMY OF THE ENDOCRINE SYSTEM

SELF REVIEW | Human Cadaver

Exercise 5.1 Hypothalamus, Pituitary Gland, and Pineal Gland

GO TO > HUMAN CADAVER > ENDOCRINE SYSTEM > SELF REVIEW > IMAGES 1 AND 2

- *Mouse over the images to locate and label the structures indicated below. Click on the structures to hear their pronunciations.*

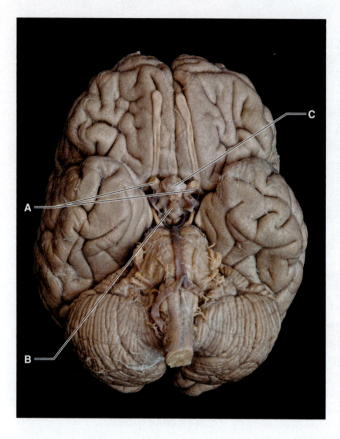

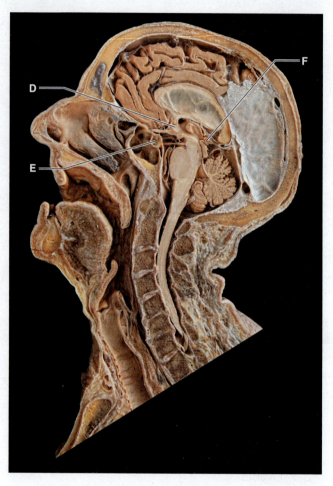

A. _____   D. _____

B. _____   E. _____

C. _____   F. _____

Exercise 5.2   **Thyroid Gland**

GO TO HUMAN CADAVER > ENDOCRINE SYSTEM > SELF REVIEW > IMAGE 4

> • *Mouse over the image to locate and label the structures indicated below. Click on the structures to hear their pronunciations.*

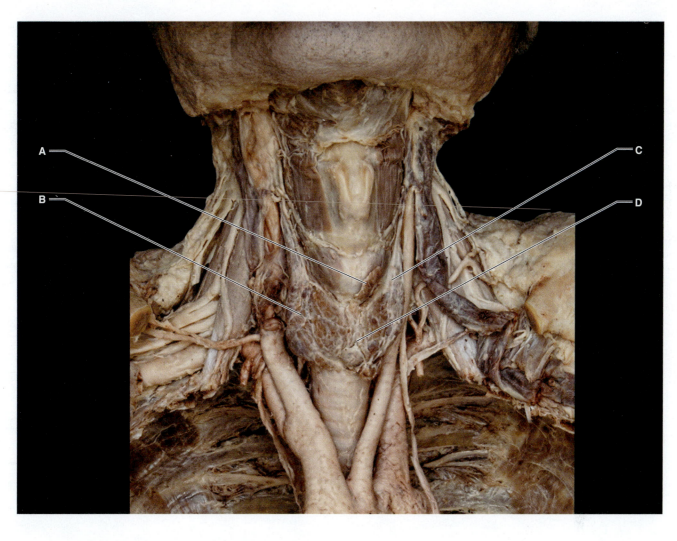

A. _____   C. _____

B. _____   D. _____

Exercise 5.3   Pancreas

GO TO ▷ HUMAN CADAVER > ENDOCRINE SYSTEM > SELF REVIEW > IMAGE 8

   • *Mouse over the image to locate and label the structures indicated below. Click on the structures to hear their pronunciations.*

A. _____   C. _____

B. _____

Exercise 5.4   **Kidneys and Adrenal Glands**

GO TO〉 HUMAN CADAVER > ENDOCRINE SYSTEM > SELF REVIEW > IMAGE 9

- *Mouse over the image to locate and label the structures indicated below. Click on the structures to hear their pronunciations. Note that the word adrenal means near to the kidney.*

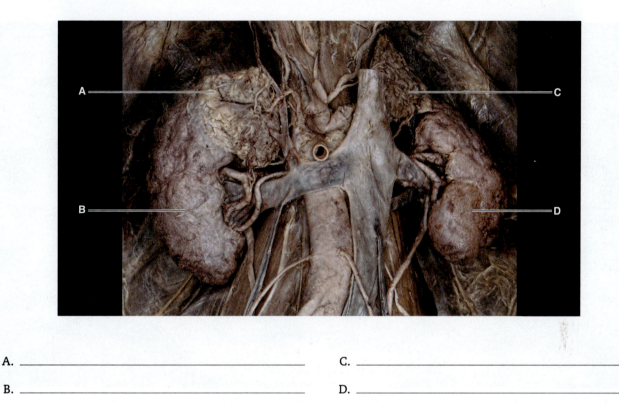

A. _____   C. _____

B. _____   D. _____

Exercise 5.5   **Testis**

GO TO〉 HUMAN CADAVER > ENDOCRINE SYSTEM > SELF REVIEW > IMAGE 11

- *Mouse over the image to locate and label the structures indicated below. Click on the structures to hear their pronunciations.*

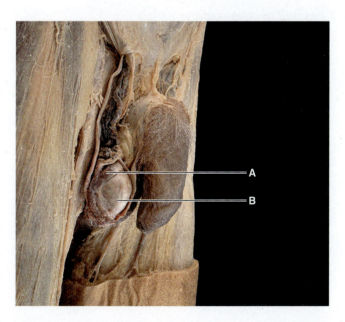

A. _____

B. _____

Exercise 5.6 Ovaries

GO TO > HUMAN CADAVER > ENDOCRINE SYSTEM > SELF REVIEW > IMAGE 12

- *Mouse over the image to locate and label the structures indicated below. Click on the structures to hear their pronunciations.*

A. _____   C. _____

B. _____   D. _____

QUIZ | Human Cadaver

Answer the following questions using information from PAL 3.1 as well as other course materials including your textbook, lecture, and lab notes.

I. Check Your Understanding

1. The thyroid gland has two lateral lobes connected anteriorly by what structure?

_____

2. The adrenal glands are located on the superior surface of which organs?

_____

3. How many parathyroid glands are typically found on the posterior surface of the thyroid?

_____

4. Which structure transports the ovulated ovum to the uterus?

_____

5. What is the name of the structure that runs from the superior surface of the testis to its posterolateral surface?

_____

_____

6. Match each organ, at left, to its correct location in the body, at right.

_____ pituitary gland   a. larynx

_____ thyroid gland   b. scrotum

_____ adrenal gland   c. retroperitoneal

_____ kidney   d. abdomen

_____ pancreas   e. pelvis

_____ ovary   f. brain

_____ testis   g. kidney

BEYOND PAL

7. Tumors of the pituitary gland are often associated with visual disturbances due to the gland's close proximity to which structure(s)?

_____

_____

8. Identify one steroid hormone produced by the ovary.

_____

_____

9. Identify the three types of steroid hormones produced by the adrenal cortex.

_____

_____

_____

10. Which endocrine gland is primarily responsible for basal metabolic rate?

_____

II. Apply What You Learned

Throughout history, castration of boys possessing exceptional voices was a common practice. Known as "castrati," these boys would never go through puberty, and consequently retained their soprano or mezzo-soprano voices.

1. During castration, which endocrine organs are removed? Which primary sex hormone is released by these organs?

_____

_____

_____

_____

2. What role does this hormone play in voice change during puberty?

_____

_____

_____

_____

_____

_____

_____

3. What other physical characteristics would you expect the "castrati" to exhibit or not exhibit?

_____

_____

_____

_____

_____

_____

_____

_____

_____

_____

SELF REVIEW | Anatomical Models

Exercise 5.7 **Hypothalamus, Pituitary Gland, and Pineal Gland**

GO TO > ANATOMICAL MODELS > ENDOCRINE SYSTEM > SELF REVIEW > IMAGES 2 AND 4

- *Mouse over the images to locate and label the structures indicated below. Click on the structures to hear their pronunciations.*

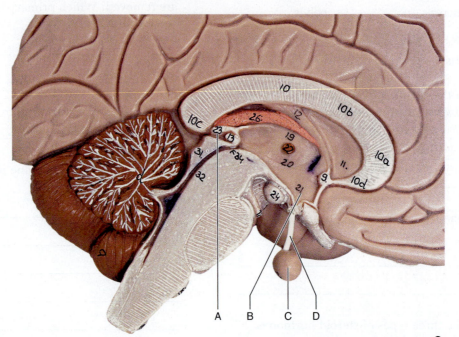

LT-C15: Brain, 2-part, 3B Scientific®

A. _____   C. _____

B. _____   D. _____

E. _____

F. _____

G. _____

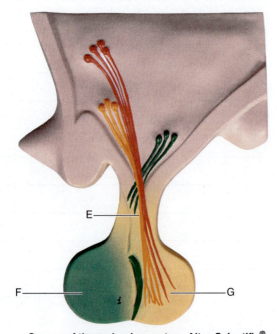

Organs of the endocrine system, Altay Scientific®, available at www.wardsci.com

Exercise 5.8   Thyroid and Parathyroid Glands

GO TO ANATOMICAL MODELS > ENDOCRINE SYSTEM > SELF REVIEW > IMAGES 8 AND 10

• *Mouse over the images to locate and label the structures indicated below. Click on the structures to hear their pronunciations.*

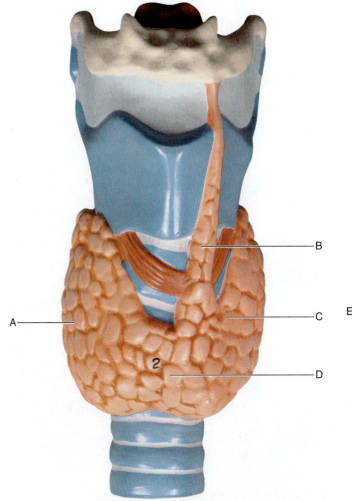

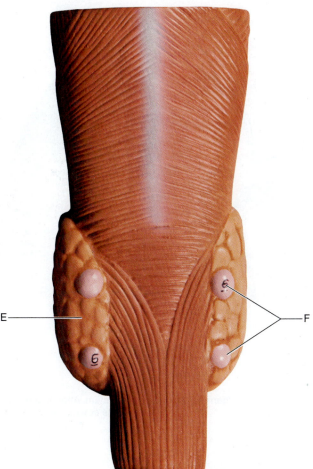

Organs of the endocrine system, Altay Scientific®, available at www.wardsci.com

Organs of the endocrine system, Altay Scientific®, available at www.wardsci.com

A. _____   D. _____

B. _____   E. _____

C. _____   F. _____

Exercise 5.9 **Adrenal Gland**

GO TO ANATOMICAL MODELS > ENDOCRINE SYSTEM > SELF REVIEW > IMAGES 11 AND 12

- *Mouse over the images to locate and label the structures indicated below. Click on the structures to hear their pronunciations. Note that the word adrenal means near to the kidney.*

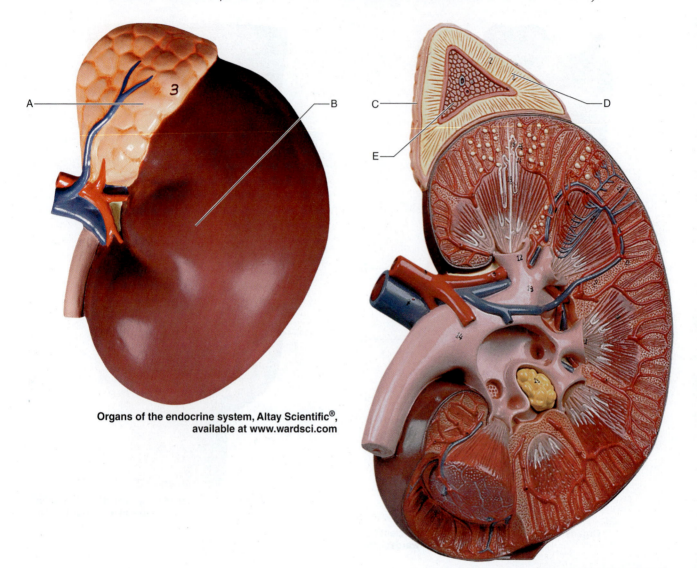

Organs of the endocrine system, Altay Scientific®, available at www.wardsci.com

Kidney with adrenal gland, Altay Scientific®, available at www.wardsci.com

A. _____

B. _____

C. _____

D. _____

E. _____

Exercise 5.10 Pancreas

GO TO ANATOMICAL MODELS > ENDOCRINE SYSTEM > SELF REVIEW > IMAGES 13 AND 15

- *Mouse over the images to locate and label the structures indicated below. Click on the structures to hear their pronunciations.*

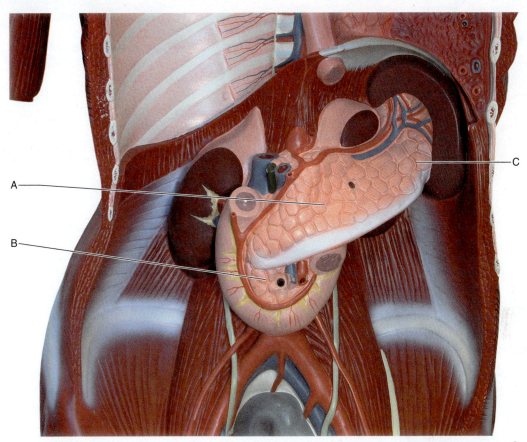

LT-VA16: LIfe-size muscle torso, 27-part, 3B Scientific®

A. _____   C. _____

B. _____

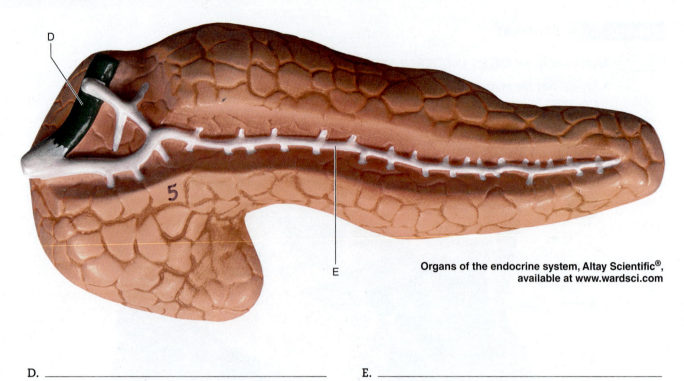

Organs of the endocrine system, Altay Scientific®,
available at www.wardsci.com

D. _____   E. _____

Exercise 5.11   Testis

GO TO ⟩ ANATOMICAL MODELS > ENDOCRINE SYSTEM > SELF REVIEW > IMAGE 17

- *Mouse over the image to locate and label the structures indicated below. Click on the structures to hear their pronunciations.*

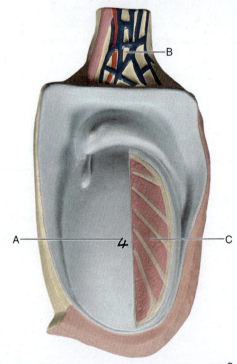

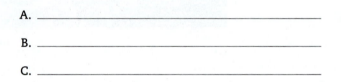

A. _____

B. _____

C. _____

Organs of the endocrine system, Altay Scientific®,
available at www.wardsci.com

Exercise 5.12   **Female Reproductive Structures**

GO TO   ANATOMICAL MODELS > ENDOCRINE SYSTEM > SELF REVIEW > IMAGES 19 AND 20

- *Mouse over the images to locate and label the structures indicated below. Click on the structures to hear their pronunciations.*

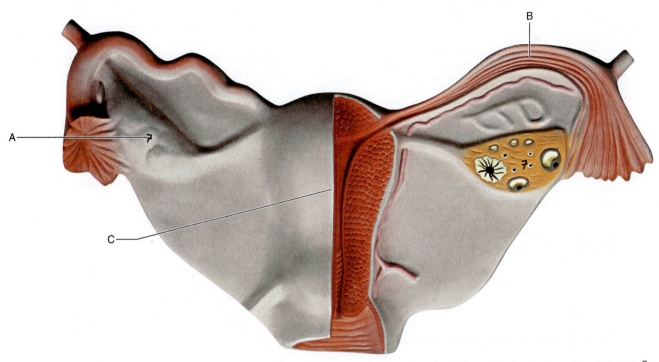

Organs of the endocrine system, Altay Scientific®, available at www.wardsci.com

A. _____

B. _____

C. _____

D. _____

E. _____

F. _____

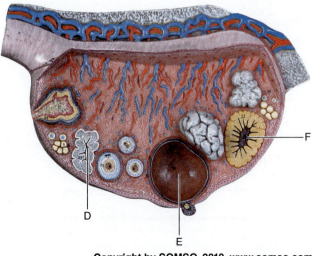

QUIZ | Anatomical Models

Answer the following questions using information from PAL 3.1 as well as other course materials including your textbook, lecture, and lab notes.

I. Check Your Understanding

1. Identify the two principal parts of the pituitary gland.

_____

_____

2. Which two nuclei of the hypothalamus synthesize hormones secreted by the posterior pituitary?

_____

_____

3. The pituitary gland is attached to the hypothalamus via what structure?

_____

_____

4. Identify the two regions of the adrenal gland.

_____

_____

5. Which endocrine glands are located on the posterior surface of the thyroid?

_____

_____

6. Identify the two micro-endocrine organs within the ovary.

_____

_____

7. What is the name of the thick connective tissue that covers the testis?

_____

BEYOND PAL **8.** Which endocrine gland plays a critical role in both our short-term and long-term responses to stress?

_____

9. Which endocrine organ is primarily responsible for the control of blood glucose?

_____

10. Which endocrine gland releases a hormone that acts to increase blood calcium?

_____

11. Match each hormone, at right, to the organ that produces it, at left. An individual organ may produce more than one hormone. However, to correctly complete this exercise, use each hormone only once.

_____ pituitary gland   a. thyroxine

_____ thyroid gland   b. testosterone

_____ adrenal gland   c. PTH

_____ parathyroid gland   d. insulin

_____ pancreas   e. estrogen

_____ ovary   f. GH

_____ testis   g. cortisol

II. Apply What You Learned

The pancreas is a major endocrine and exocrine organ and, if damaged, can cause many health problems and potentially death.

1. What is the major endocrine function of the pancreas?

_____

_____

_____

_____

2. How do the endocrine secretions move out of the pancreas?

_____

_____

_____

_____

3. Type 1 diabetes is caused by a reduction in a specific hormone-producing cell of the pancreas. It is thought that these cells are attacked by the individual's own immune system. Which cells are attacked, and what is the result of a decrease in these cells?

_____

_____

_____

_____

_____

_____

_____

4. What is the exocrine function of the pancreas?

_____

_____

_____

_____

_____

_____

5. To which organ does the pancreas have a direct connection? What is transported to this organ via a system of ducts?

_____

_____

_____

_____

_____

TISSUES OF THE ENDOCRINE SYSTEM

SELF REVIEW | Histology

Exercise 5.13 **Pituitary Gland, Cross Section 40x**

GO TO › HISTOLOGY > ENDOCRINE SYSTEM > SELF REVIEW > IMAGE 1

- *Mouse over the image to locate and label the structures indicated below. Click on the structures to hear their pronunciations.*

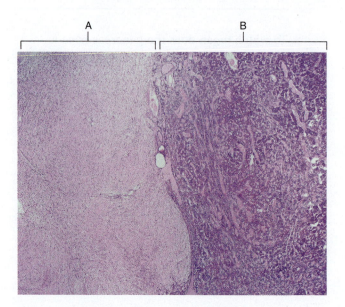

A. _____

B. _____

Exercise 5.14 **Thyroid Gland, Cross Section 400x**

GO TO › HISTOLOGY > ENDOCRINE SYSTEM > SELF REVIEW > IMAGE 5

- *Mouse over the image to locate and label the structures indicated below. Click on the structures to hear their pronunciations.*

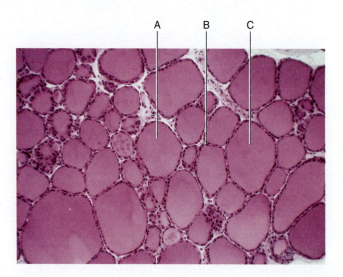

A. (structure) _____

B. _____

C. (material) _____

Exercise 5.15   **Thyroid and Parathyroid Glands, Cross Section 100x**

GO TO HISTOLOGY > ENDOCRINE SYSTEM > SELF REVIEW > IMAGE 8

- *Mouse over the image to locate and label the structures indicated below. Click on the structures to hear their pronunciations.*

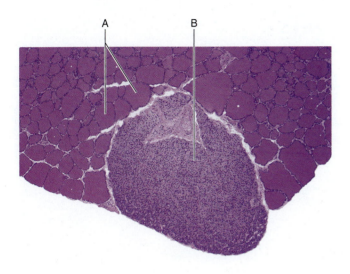

A. (structures) _____

B. _____

Exercise 5.16   **Adrenal Gland, Cross Section 100x**

GO TO HISTOLOGY > ENDOCRINE SYSTEM > SELF REVIEW > IMAGE 13

- *Mouse over the image to locate and label the structures indicated below. Click on the structures to hear their pronunciations.*

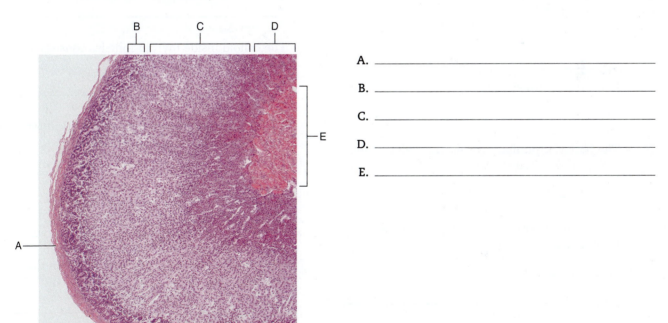

A. _____

B. _____

C. _____

D. _____

E. _____

Exercise 5.17 **Pancreas, Cross Section 100x**

GO TO HISTOLOGY > ENDOCRINE SYSTEM > SELF REVIEW > IMAGE 16

- *Mouse over the image to locate and label the structures indicated below. Click on the structures to hear their pronunciations.*

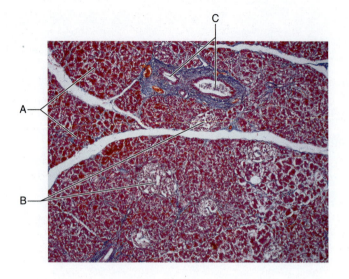

A. _____

B. _____

C. _____

QUIZ | Histology

Answer the following questions using information from PAL 3.1 as well as other course materials including your textbook, lecture, and lab notes.

I. Check Your Understanding

1. Which endocrine gland consists of spherical sacs lined by simple cuboidal epithelium that contain colloid?

_____

2. The anterior pituitary consists of which of the four primary tissue types?

_____

3. The posterior pituitary consists of which of the four primary tissue types?

_____

4. Identify the hormone-secreting structures located within the pancreas.

_____

_____

5. Which endocrine gland consists of an inner medulla and an outer cortex, all of which is surrounded by a connective tissue capsule?

_____

BEYOND PAL **6.** The colloid inside thyroid follicles consists of what large protein?

_____

7. Which adrenal hormone(s) is (are) critical for our short-term response to stress?

_____

_____

8. The thyroid gland is unique because it stores a significant amount of hormone. In what form are thyroid hormones stored within the thyroid gland?

_____

_____

9. The pituitary gland is considered part of the diencephalon, and is closely controlled by what other part of the diencephalon?

_____

_____

10. Identify the hormone that is released by chief cells of the parathyroid gland.

_____

11. The following is a list of three different chemical classes of hormones.

a. amines–derivatives of one or two amino acids

b. proteins–chains of amino acids

c. steroids–derivatives of cholesterol

Identify which class of hormone is secreted by each of the following endocrine glands.

_____ anterior pituitary

_____ adrenal cortex

_____ thyroid gland

_____ pancreas

_____ adrenal medulla

_____ parathyroid gland

II. Apply What You Learned

Four small glands, called parathyroid glands, are located on the posterior surface of the thyroid gland or even within the thyroid tissue itself. Not described until 1850, these small structures were the last organs to be discovered. Their small size and location led to many deaths during surgical procedures involving the thyroid gland as the true importance of their function was not truly understood at the time.

1. What is the function of the parathyroid glands?

_____

_____

2. What hormone, released by these glands, is responsible for its regulatory role in the body? How does this hormone exert its effects?

_____

_____

_____

_____

_____

_____

_____

_____

LAB PRACTICAL Endocrine System

1. Identify the organ in the image at right.

   _____

2. Identify the structures.

   _____

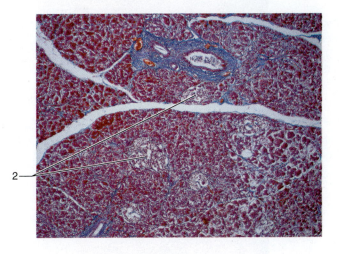

3. Identify the organ in the image at right.

   _____

4. Identify the structure.

   _____

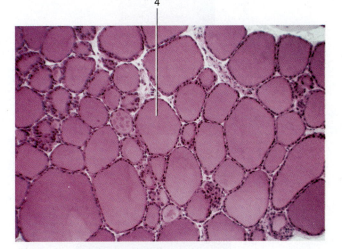

LAB PRACTICAL *continues*

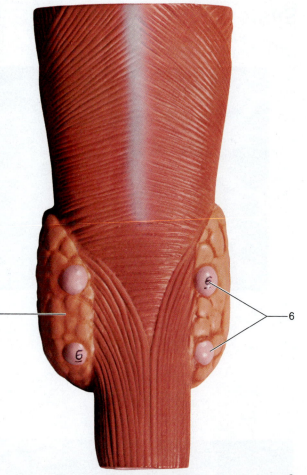

Organs of the endocrine system, Altay Scientific®, available at www.wardsci.com

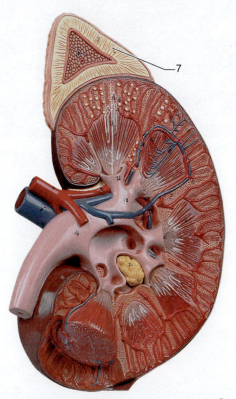

Kidney with adrenal gland, Altay Scientific®, available at www.wardsci.com

5. Identify the gland.

   _____

6. Identify the glands.

   _____

7. Identify the gland.

   _____

8. Which region of this gland secretes catecholamines?

   _____

9. Identify the organ.

_____

10. What are the two main steroid hormones produced by this organ?

_____

11. Identify the gland in the image at right.

_____

12. Identify the specific zone of this gland.

_____

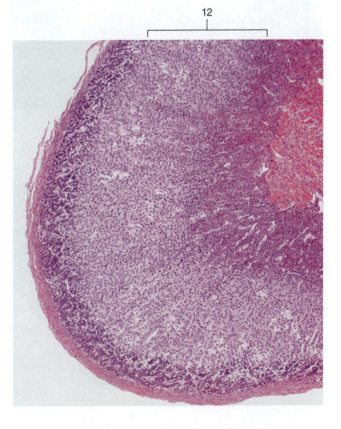

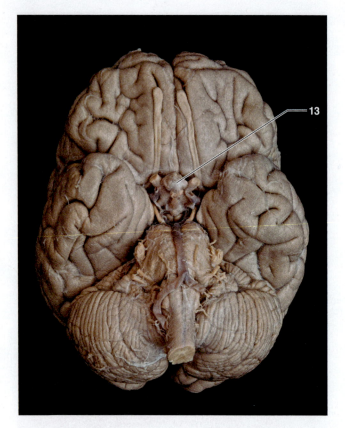

13. Identify the gland.

_____

14. Which structure controls hormone secretion by this gland?

_____

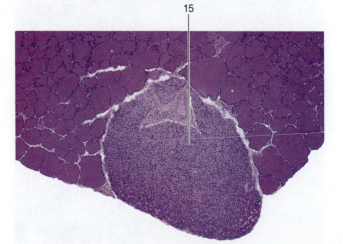

15. Identify the gland.

_____

16. Identify the hormone this gland secretes.

_____

17. Identify the organ.

_____

18. Identify the hormone released by the alpha-cells of this organ.

_____

19. Identify the gland in the image at right.

_____

20. Identify the specific region of this gland.

_____

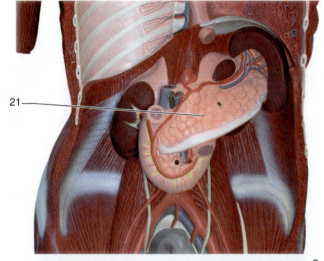

21. Identify the specific region of the organ.

_____

22. This organ consists of two types of glandular epithelia, which are named according to where their products are released. Identify these two types of glandular epithelia.

_____

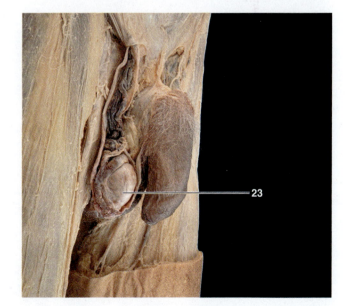

LT-VA16: LIfe-size muscle torso, 27-part, 3B Scientific®

23. Identify the organ.

_____

24. Which type of steroid does this organ secrete?

_____

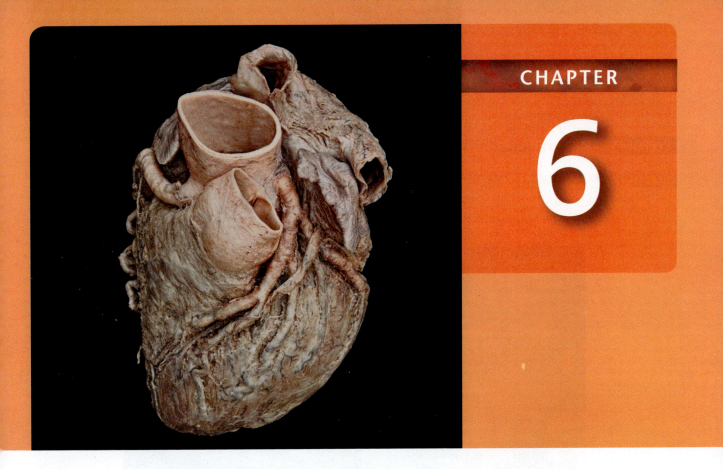

The Cardiovascular System

STUDENT OBJECTIVES

HEART

1. Understand the important structural features of each heart chamber: right and left atria, and right and left ventricles.
2. Name the heart valves and describe their locations and functions.
3. Describe the path of a drop of blood through the four chambers of the heart and the pulmonary and systemic circuits.
4. Describe the locations of the coronary arteries and cardiac veins on the heart surface.

ARTERIES

5. Describe the path of a drop of blood as it passes through the pulmonary and systemic circuits, starting at the right atrium of the heart.
6. Identify the major arteries of the pulmonary and systemic circuits. Describe their locations and determine the body regions they supply.

VEINS

7. Identify the major veins of the pulmonary and systemic circuits. Describe their locations and determine the body regions they drain.

TISSUES OF THE CARDIOVASCULAR SYSTEM

8. Describe the structure of cardiac muscle tissue.
9. Describe the three tunics that form the wall of an artery or vein.

HEART

SELF REVIEW | Human Cadaver

Exercise 6.1 **Heart, Anterior View**

GO TO ⟩ HUMAN CADAVER > CARDIOVASCULAR SYSTEM > HEART > SELF REVIEW > IMAGE 5

- *Mouse over the image to locate and label the structures indicated below. Click on the structures to hear their pronunciations.*

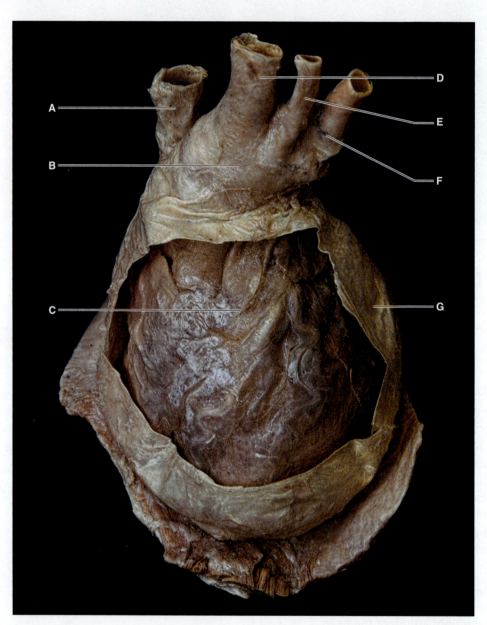

A. _____   E. _____

B. _____   F. _____

C. _____   G. _____

D. _____

Exercise 6.2   Heart, Superior View

GO TO ⟩ HUMAN CADAVER > CARDIOVASCULAR SYSTEM > HEART > SELF REVIEW > IMAGE 8

- *Mouse over the image to locate and label the structures indicated below. Click on the structures to hear their pronunciations.*

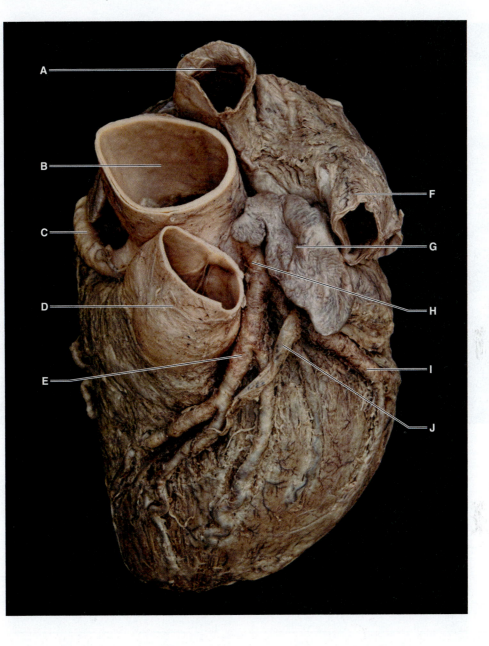

A. _____

B. _____

C. _____

D. _____

E. _____

F. _____

G. _____

H. _____

I. _____

J. _____

Exercise 6.3   **Heart Structures**

GO TO ▷ HUMAN CADAVER > CARDIOVASCULAR SYSTEM > HEART > SELF REVIEW > IMAGES 10 AND 11

• *Mouse over the images to locate and label the structures indicated below. Click on the structures to hear their pronunciations.*

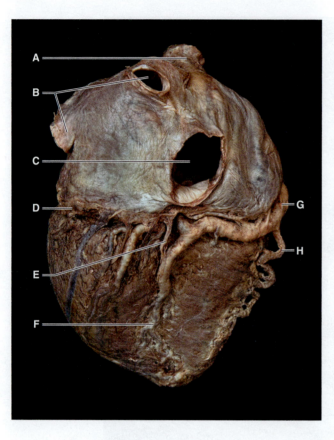

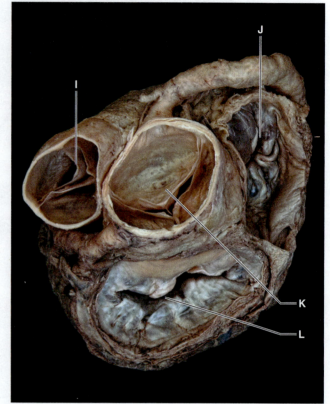

A. _____

B. _____

C. _____

D. _____

E. _____

F. _____

G. _____

H. _____

I. _____

J. _____

K. _____

L. _____

Exercise 6.4   **Heart, Internal View**

GO TO ⟩ HUMAN CADAVER > CARDIOVASCULAR SYSTEM > HEART > SELF REVIEW > IMAGE 14

- *Mouse over the image to locate and label the structures indicated below. Click on the structures to hear their pronunciations.*

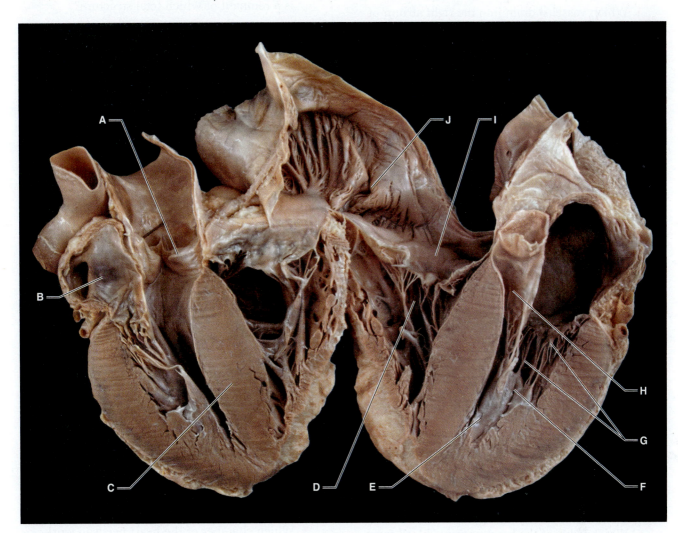

A. _____

B. (chamber) _____

C. _____

D. (chamber) _____

E. _____

F. (chamber) _____

G. _____

H. _____

I. _____

J. (chamber) _____

QUIZ) Human Cadaver

Answer the following questions using information from PAL 3.1 as well as other course materials including your textbook, lecture, and lab notes.

I. Check Your Understanding

1. Which vessel(s) drain into the right atrium of the heart?

   _____

   _____

   _____

2. Which chamber(s) of the heart contain the chordae tendineae?

   _____

   _____

3. Blood traveling through the pulmonary trunk will then pass into what organ(s)?

   _____

   _____

4. The fossa ovalis is found on which septum of the heart?

   _____

5. Which blood vessel travels through the posterior interventricular sulcus, draining blood from the posterior surface of the heart?

   _____

6. Which valve of the heart prevents blood from flowing back into the right ventricle?

   _____

7. Which chamber of the heart has the thickest layer of myocardium?

   _____

BEYOND PAL 8. Deoxygenated blood is found in which chamber(s) of the heart?

   _____

   _____

9. The fossa ovalis is found in the adult heart and is a remnant of which fetal structure?

   _____

10. The endocardium of the heart is composed of what type of epithelium?

   _____

II. Apply What You Learned

Aortic valve stenosis is one condition that results in a heart murmur. This condition can develop in adults several years after they have had rheumatic fever as a result of scar tissue forming from the infection. It may also develop due to calcium deposits building around the cusps of the valve. The result is a narrowing of the valve opening.

1. Which valve of the heart is affected in aortic valve stenosis?

   _____

2. How does narrowing of this heart valve affect blood flow to the tissues of the body?

   _____

   _____

   _____

   _____

3. Which chamber of the heart needs to work harder to accommodate for the narrowed valve?

   _____

   _____

   _____

   _____

4. What happens to the heart chamber wall as a result of this increased workload?

   _____

   _____

   _____

   _____

SELF REVIEW | Anatomical Models

Exercise 6.5 **Heart, Anterior View**

GO TO ANATOMICAL MODELS > CARDIOVASCULAR SYSTEM > HEART > SELF REVIEW > IMAGES 1 AND 2

- *Mouse over the images to locate and label the structures indicated below. Click on the structures to hear their pronunciations.*

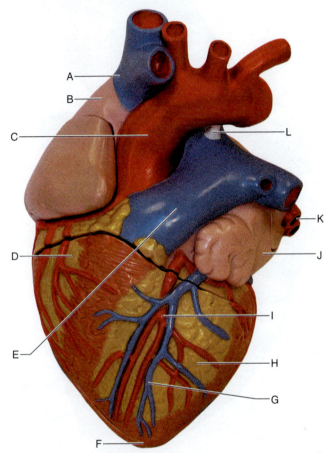

LT-G12: Giant heart, 4-part, 3B Scientific®

A. _____   G. _____

B. (chamber) _____   H. (chamber) _____

C. _____   I. _____

D. (chamber) _____   J. (chamber) _____

E. _____   K. _____

F. _____   L. _____

Exercise 6.6 **Heart, Internal Structures**

GO TO ▷ ANATOMICAL MODELS > CARDIOVASCULAR SYSTEM > HEART > SELF REVIEW > IMAGE 8

- *Mouse over the image to locate and label the structures indicated below. Click on the structures to hear their pronunciations.*

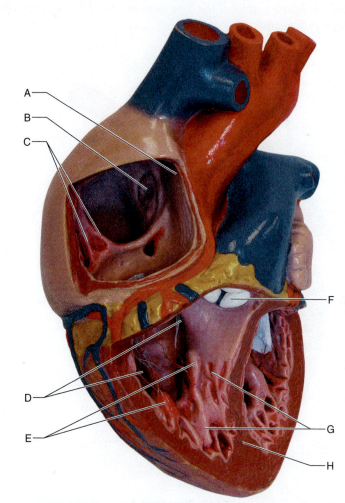

LT-G12: Giant heart, 4-part, 3B Scientific®

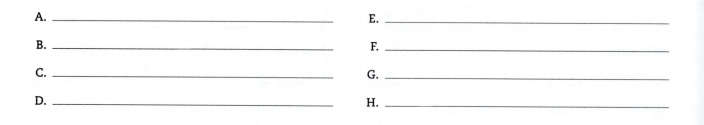

A. _____   E. _____

B. _____   F. _____

C. _____   G. _____

D. _____   H. _____

Exercise 6.7   Heart, Posterior View

GO TO ANATOMICAL MODELS > CARDIOVASCULAR SYSTEM > HEART > SELF REVIEW > IMAGES 5 AND 6

• *Mouse over the image to locate and label the structures indicated below. Click on the structures to hear their pronunciations.*

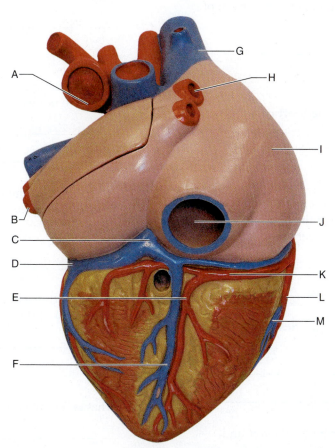

LT-G12: Giant heart, 4-part, 3B Scientific®

A. _____

B. _____

C. _____

D. _____

E. _____

F. _____

G. _____

H. _____

I. (chamber) _____

J. _____

K. _____

L. _____

M. _____

QUIZ | Anatomical Models

Answer the following questions using information from PAL 3.1 as well as other course materials including your textbook, lecture, and lab notes.

I. Check Your Understanding

1. Which vessel(s) drain into the left atrium of the heart?

   _____

   _____

2. Which chamber(s) of the heart contain papillary muscles?

   _____

   _____

3. Blood traveling through the bicuspid valve is flowing into what chamber or vessel?

   _____

   _____

4. What are the first two arteries that branch from the aorta?

   _____

   _____

5. Within which chambers would you find pectinate muscles?

   _____

   _____

6. Which heart sulcus contains the great cardiac vein?

   _____

   _____

7. Which structures of the heart wall anchor the chordae tendineae?

   _____

   _____

8. Which valve of the heart separates the left ventricle from the aorta?

   _____

   _____

BEYOND PAL 9. Oxygenated blood is found in which chamber(s) of the heart?

   _____

   _____

10. What is the function of the tricuspid valve?

   _____

   _____

   _____

   _____

II. Apply What You Learned

A myocardial infarction, commonly known as a heart attack, occurs when heart muscle is either damaged or dead as a result of a loss of an oxygenated blood supply. This is often caused by a blockage in one or more of the blood vessels supplying the heart wall.

1. What are the two primary vessels that branch off the aorta to supply blood to the heart?

   _____

   _____

2. If a blockage occurs in the posterior interventricular artery, one consequence is that the blood supply to the posteromedial papillary muscle of the left ventricle is disrupted. The function of which heart valve will be affected by damage to this muscle? What will be the result of damage to this valve?

   _____

   _____

   _____

   _____

   _____

3. Damage to the myocardium of the left ventricle also impacts the force with which the chamber can pump blood. How does this damage impact heart function?

   _____

   _____

   _____

   _____

   _____

   _____

ARTERIES

SELF REVIEW | Human Cadaver

Exercise 6.8 Arteries of the Brain, Inferior View

GO TO > HUMAN CADAVER > CARDIOVASCULAR SYSTEM > BLOOD VESSELS > SELF REVIEW > IMAGE 9

- *Mouse over the image to locate and label the vessels indicated below. Click on the vessels to hear their pronunciations.*

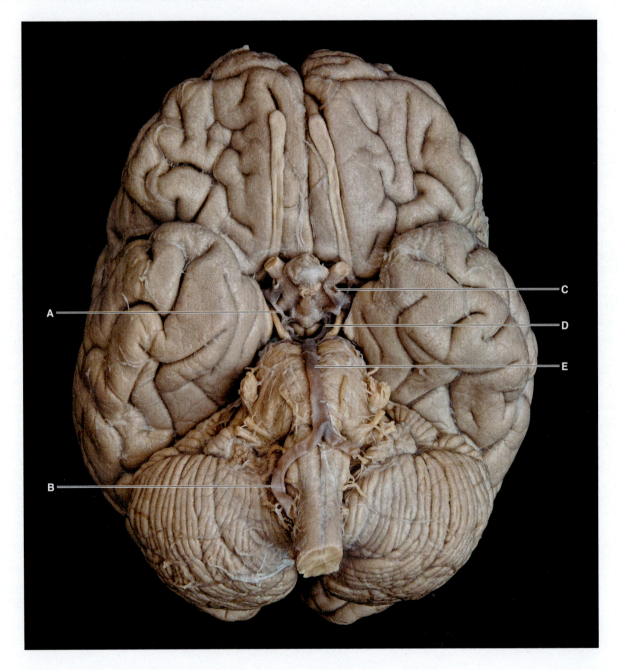

A. _____

B. _____

C. _____

D. _____

E. _____

Exercise 6.9 Arteries of the Thorax, Anterior View

GO TO HUMAN CADAVER > CARDIOVASCULAR SYSTEM > BLOOD VESSELS > SELF REVIEW > IMAGE 14

- *Mouse over the image to locate and label the vessels indicated below. Click on the vessels to hear their pronunciations.*

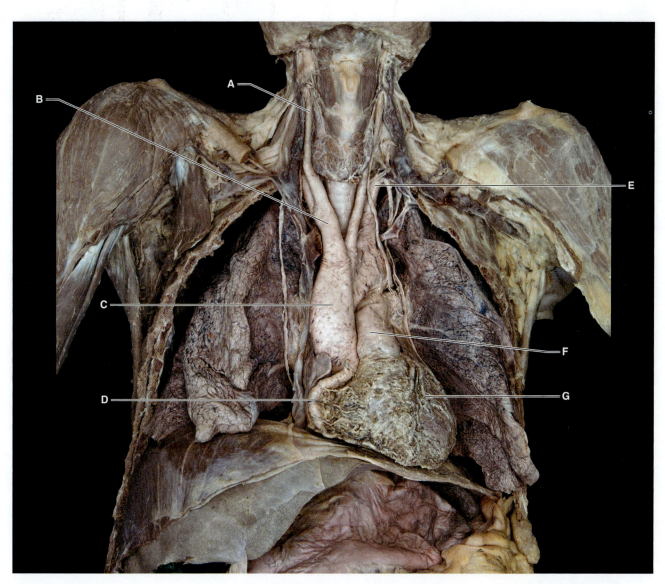

A. _____   E. _____

B. _____   F. _____

C. _____   G. _____

D. _____

Exercise 6.10 **Arteries of the Abdominal Cavity**

GO TO HUMAN CADAVER > CARDIOVASCULAR SYSTEM > BLOOD VESSELS > SELF REVIEW > IMAGE 22

- *Mouse over the image to locate and label the vessels indicated below. Click on the vessels to hear their pronunciations.*

A. _____   D. _____

B. _____   E. _____

C. _____

Exercise 6.11   Major Arteries

GO TO ⟩ HUMAN CADAVER > CARDIOVASCULAR SYSTEM > BLOOD VESSELS > SELF REVIEW > IMAGE 23

- *Mouse over the image to locate and label the vessels indicated below. Click on the vessels to hear their pronunciations.*

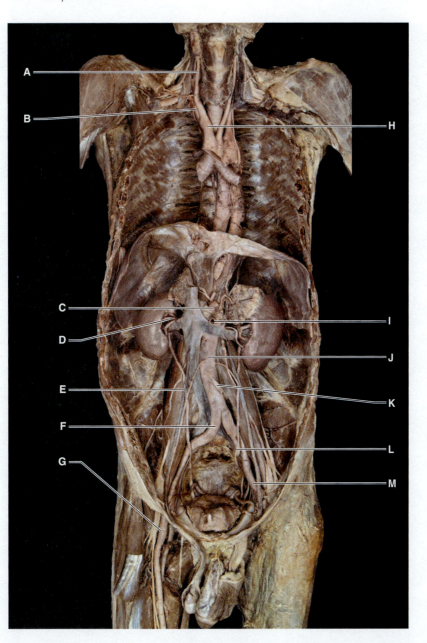

A. _____   H. _____

B. _____   I. _____

C. _____   J. _____

D. _____   K. _____

E. _____   L. _____

F. _____   M. _____

G. _____

Exercise 6.12   Arteries of the Upper Limb and Neck

GO TO ▷ HUMAN CADAVER > CARDIOVASCULAR SYSTEM > BLOOD VESSELS > SELF REVIEW > IMAGE 31

- *Mouse over the image to locate and label the vessels indicated below. Click on the vessels to hear their pronunciations.*

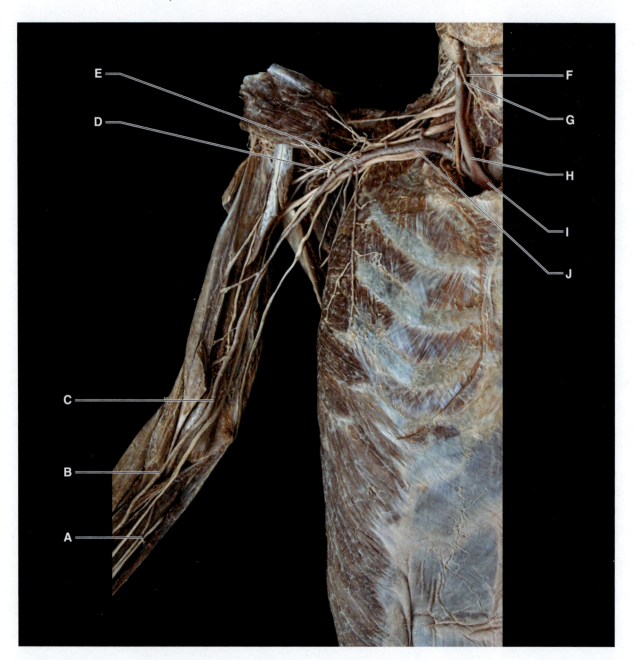

A. _____	F. _____
B. _____	G. _____
C. _____	H. _____
D. _____	I. _____
E. _____	J. _____

Exercise 6.13   Arteries of the Lower Limb

GO TO › HUMAN CADAVER > CARDIOVASCULAR SYSTEM > BLOOD VESSELS > SELF REVIEW > IMAGES 36 AND 37

- *Mouse over the images to locate and label the vessels indicated below. Click on the vessels to hear their pronunciations.*

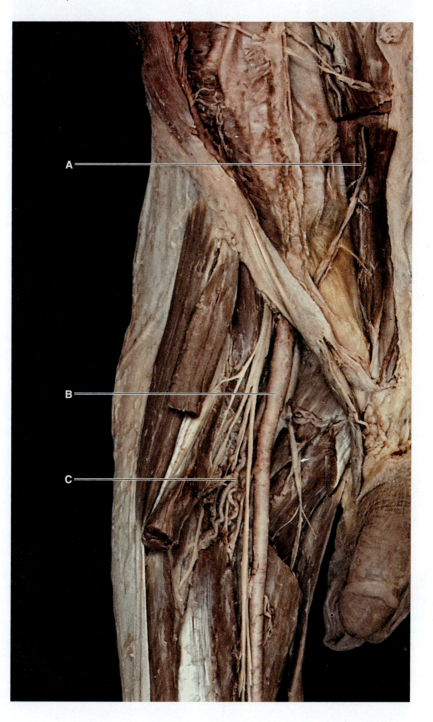

A. _____   C. _____

B. _____

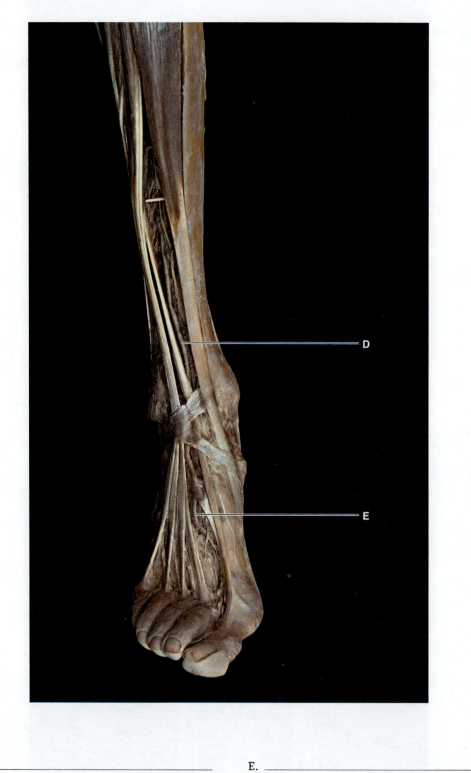

D. _____   E. _____

QUIZ | Human Cadaver

Answer the following questions using information from PAL 3.1 as well as other course materials including your textbook, lecture, and lab notes.

I. Check Your Understanding

1. What are the three major arteries that typically branch from the aortic arch?

   _____

   _____

   _____

2. The terminal end of the aorta branches into which paired vessels?

   _____

   _____

   _____

3. What are the three branches of the celiac trunk?

   _____

   _____

   _____

4. The brachial artery is a branch of what vessel?

   _____

   _____

5. Which blood vessel supplies the kidney?

   _____

   _____

6. From which vessel does the anterior tibial artery branch?

   _____

   _____

7. Which branch of the aorta is the primary blood supply for the ileum and jejunum?

   _____

   _____

8. Which paired artery passes through the carotid canal, supplying blood to the brain?

   _____

   _____

BEYOND PAL 9. What is the definition of an artery?

   _____

   _____

10. Is the aorta part of the pulmonary or systemic circuit?

   _____

   _____

II. Apply What You Learned

A stroke occurs when the blood flow to a region of the brain is disrupted or stopped. It can occur along any of the vessels that supply blood to the brain, and brain damage happens quickly. An ischemic stroke occurs when there is a blockage in a blood vessel of the brain. A hemorrhagic stroke occurs when a weak blood vessel bursts.

1. Which four major vessels supply the brain with blood?

   _____

   _____

   _____

   _____

2. Which of the following vessels that make up the arterial circle of Willis is *not* paired?

   a. anterior cerebral
   b. anterior communicating
   c. middle cerebral
   d. posterior cerebral
   e. posterior communicating

3. How might the circle of Willis prevent a stroke from causing more damage?

   _____

   _____

   _____

   _____

   _____

   _____

4. Which type of stroke would respond best to treatment that involved the use of blood thinners or thrombolytic therapy? Why?

   _____

   _____

   _____

   _____

   _____

   _____

SELF REVIEW | Anatomical Models

Exercise 6.14   Arteries of the Thoracic Cavity

GO TO ANATOMICAL MODELS > CARDIOVASCULAR SYSTEM > ARTERIES > SELF REVIEW > IMAGE 9

- *Mouse over the image to locate and label the vessels indicated below. Click on the vessels to hear their pronunciations.*

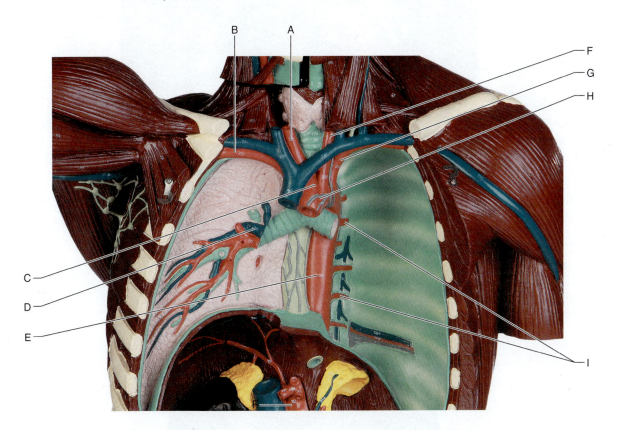

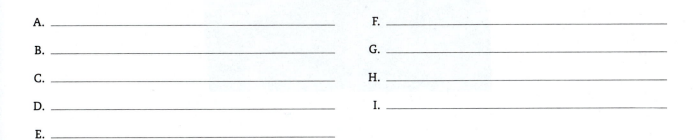

A. _____   F. _____

B. _____   G. _____

C. _____   H. _____

D. _____   I. _____

E. _____

Exercise 6.15 **Arteries of the Abdominal Cavity**

GO TO ⟩ ANATOMICAL MODELS > CARDIOVASCULAR SYSTEM > ARTERIES > SELF REVIEW > IMAGE 13

- *Mouse over the image to locate and label the vessels indicated below. Click on the vessels to hear their pronunciations.*

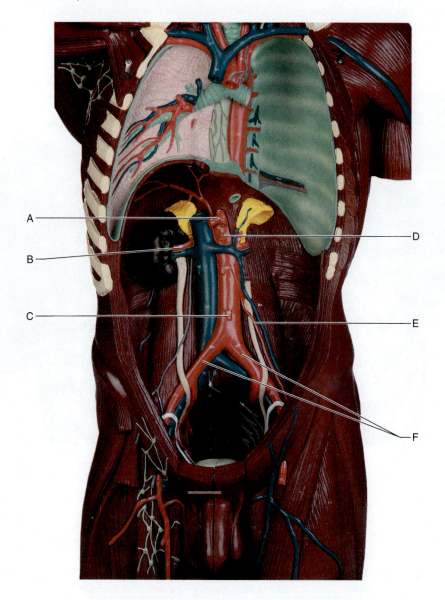

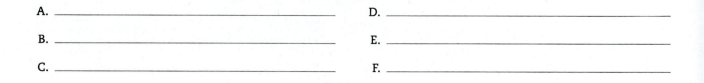

A. _____   D. _____

B. _____   E. _____

C. _____   F. _____

Exercise 6.16   Arteries of the Upper Limb and Neck

GO TO〉 ANATOMICAL MODELS > CARDIOVASCULAR SYSTEM > ARTERIES > SELF REVIEW > IMAGE 15

- *Mouse over the image to locate and label the vessels indicated below. Click on the vessels to hear their pronunciations.*

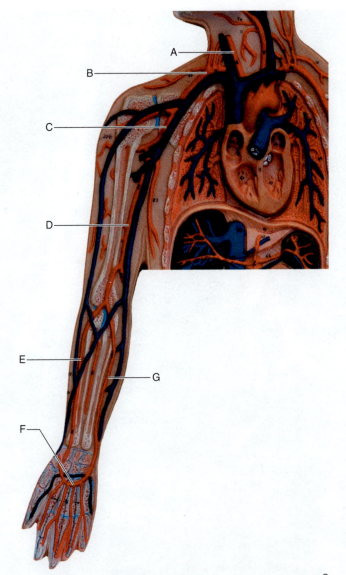

LT-G30: Human circulatory system, 3B Scientific®

A. _____   E. _____

B. _____   F. _____

C. _____   G. _____

D. _____

Exercise 6.17   Arteries of the Pelvis

GO TO ⟩ ANATOMICAL MODELS > CARDIOVASCULAR SYSTEM > ARTERIES > SELF REVIEW > IMAGE 14

- *Mouse over the image to locate and label the vessels indicated below. Click on the vessels to hear their pronunciations.*

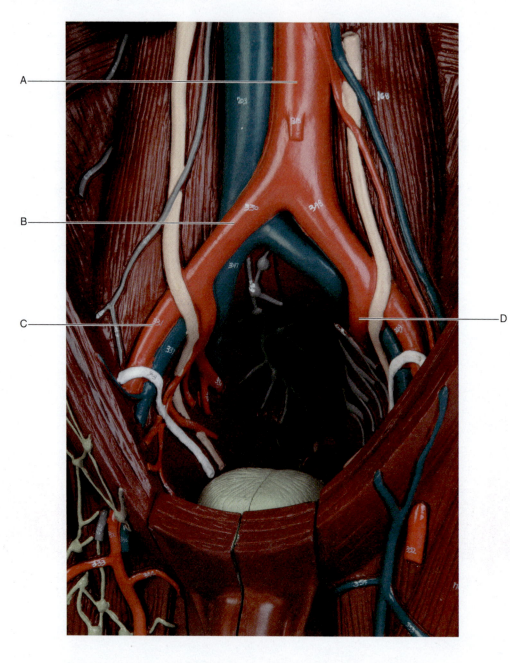

A. _____   C. _____

B. _____   D. _____

Exercise 6.18 **Arteries of the Lower Limb**

GO TO ANATOMICAL MODELS > CARDIOVASCULAR SYSTEM > ARTERIES > SELF REVIEW > IMAGE 18

- *Mouse over the image to locate and label the vessels indicated below. Click on the vessels to hear their pronunciations.*

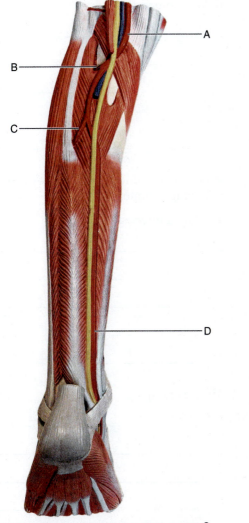

A. _____

B. _____

C. _____

D. _____

LT-M21: Muscular leg, 7-part, 3B Scientific®

QUIZ Anatomical Models

Answer the following questions using information from PAL 3.1 as well as other course materials including your textbook, lecture, and lab notes.

I. Check Your Understanding

1. Into which vessels does the brachiocephalic trunk branch?

_____

_____

2. The superficial palmar arch is primarily formed by which artery?

_____

3. Which branch of the celiac trunk supplies the liver with oxygen-rich blood?

_____

4. The femoral artery is a branch of which vessel?

_____

5. What are the terminal branches of the brachial artery?

_____

_____

6. The popliteal artery branches into which vessels?

_____

_____

BEYOND PAL 7. Which vessel branching off the subclavian artery supplies the brain with blood?

_____

_____

8. Which vessel that branches off the aorta supplies the descending colon?

_____

_____

9. In addition to the spleen, the branches of the splenic artery supply what other organ(s)?

_____

_____

10. Which arteries carry deoxygenated blood?

_____

_____

II. Apply What You Learned

Peripheral arterial disease (PAD) is characterized by tingling or numbness of the limbs, muscle cramping, and/or cold extremities. It is most frequently caused by reduced blood flow to the region.

1. What would be the most likely cause of reduced blood flow?

_____

_____

_____

_____

_____

2. Why would reduced blood flow cause the following symptoms?

Tingling/numbness

_____

_____

_____

_____

Muscle cramping

_____

_____

_____

_____

Cold extremities

_____

_____

_____

_____

_____

VEINS

SELF REVIEW | Human Cadaver

Exercise 6.19   Veins of the Head and Neck

GO TO > HUMAN CADAVER > CARDIOVASCULAR SYSTEM > BLOOD VESSELS > SELF REVIEW > IMAGE 1

- *Mouse over the image to locate and label the vessels indicated below. Click on the vessels to hear their pronunciations.*

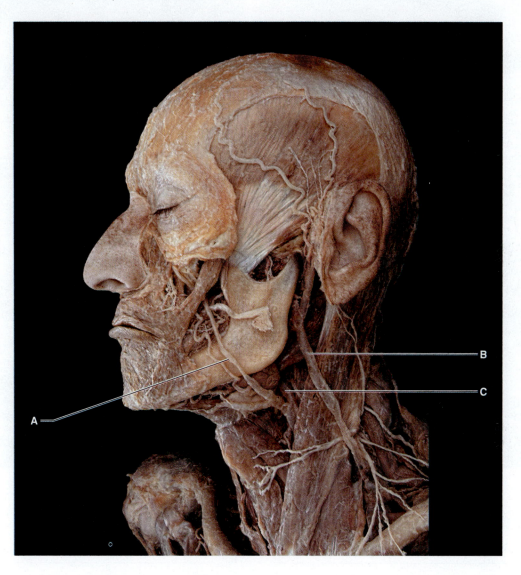

A. _____   C. _____

B. _____

Exercise 6.20   **Veins of the Thoracic Cavity**

GO TO HUMAN CADAVER > CARDIOVASCULAR SYSTEM > BLOOD VESSELS > SELF REVIEW > IMAGE 13

- *Mouse over the image to locate and label the vessels indicated below. Click on the vessels to hear their pronunciations.*

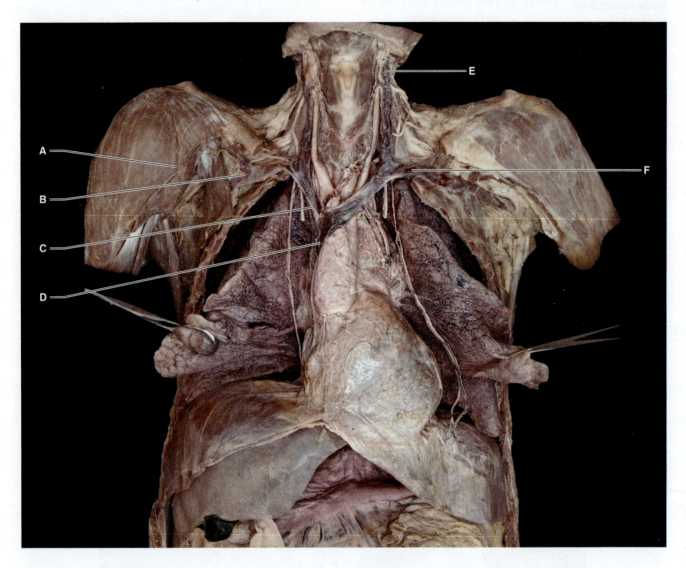

A. _____

B. _____

C. _____

D. _____

E. _____

F. _____

Exercise 6.21 Veins of the Upper and Lower Limbs

GO TO > HUMAN CADAVER > CARDIOVASCULAR SYSTEM > BLOOD VESSELS > SELF REVIEW > IMAGES 27 AND 33

- *Mouse over the images to locate and label the vessels indicated below. Click on the vessels to hear their pronunciations.*

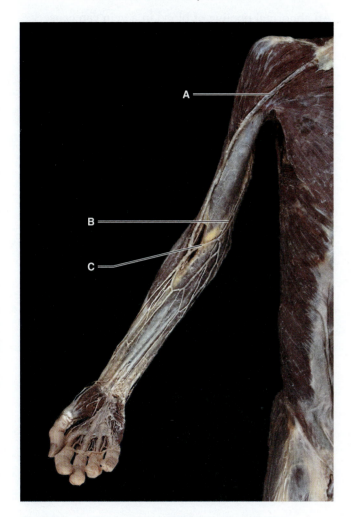

A. _____   D. _____

B. _____   E. _____

C. _____   F. _____

QUIZ | Human Cadaver

Answer the following questions using information from PAL 3.1 as well as other course materials including your textbook, lecture, and lab notes.

I. Check Your Understanding

1. The axillary vein drains into which larger vessel?

   _____

   _____

2. What does the superior vena cava drain into?

   _____

   _____

3. Which vein, commonly used for drawing blood, connects the cephalic and basilic veins?

   _____

   _____

BEYOND PAL 4. What is the definition of a vein?

   _____

   _____

5. The internal jugular vein exits the skull through which foramen?

   _____

   _____

6. Which vein is part of the femoral triangle?

   _____

   _____

II. Apply What You Learned

An abnormal pooling of blood in a vein can cause swelling and twisting of the veins. These visible and sometimes painful vessels are called varicose veins and are most commonly found in the lower limbs.

1. Which structures of a vein are not functioning properly when blood is allowed to pool instead of move back toward the heart?

   _____

   _____

2. Varicose veins tend to be aggravated by pregnancy and/or standing for long periods of time. Why might both of these circumstances contribute to the condition?

   _____

   _____

   _____

   _____

   _____

   _____

   _____

   _____

   _____

   _____

   _____

SELF REVIEW | Anatomical Models

Exercise 6.22 Dural Sinuses and Veins of the Head

GO TO > ANATOMICAL MODELS > CARDIOVASCULAR SYSTEM > VEINS > SELF REVIEW > IMAGES 2 AND 4

- *Mouse over the images to locate and label the vessels indicated below. Click on the vessels to hear their pronunciations.*

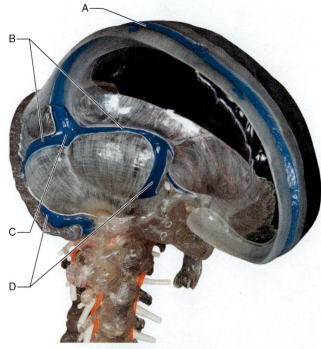

Copyright by SOMSO, 2010, www.somso.com

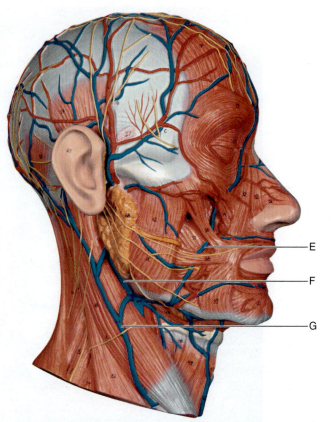

LT-VB126: Head musculature with blood vessels, 3B Scientific®

A. _____   E. _____

B. _____   F. _____

C. _____   G. _____

D. _____

Exercise 6.23 Veins of the Thorax and Upper Limb

GO TO ANATOMICAL MODELS > CARDIOVASCULAR SYSTEM > VEINS > SELF REVIEW > IMAGE 10

- *Mouse over the image to locate and label the vessels indicated below. Click on the vessels to hear their pronunciations.*

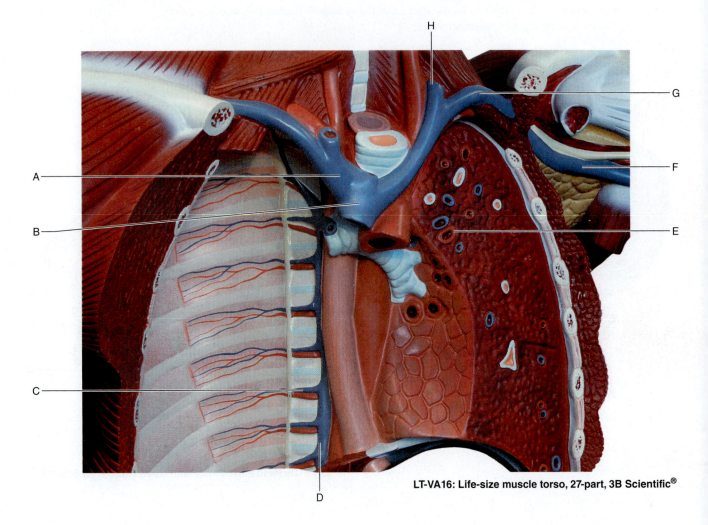

LT-VA16: Life-size muscle torso, 27-part, 3B Scientific®

A. _____   E. _____

B. _____   F. _____

C. _____   G. _____

D. _____   H. _____

Exercise 6.24 Major Veins

GO TO ANATOMICAL MODELS > CARDIOVASCULAR SYSTEM > VEINS > SELF REVIEW > IMAGE 9

- *Mouse over the image to locate and label the vessels indicated below. Click on the vessels to hear their pronunciations.*

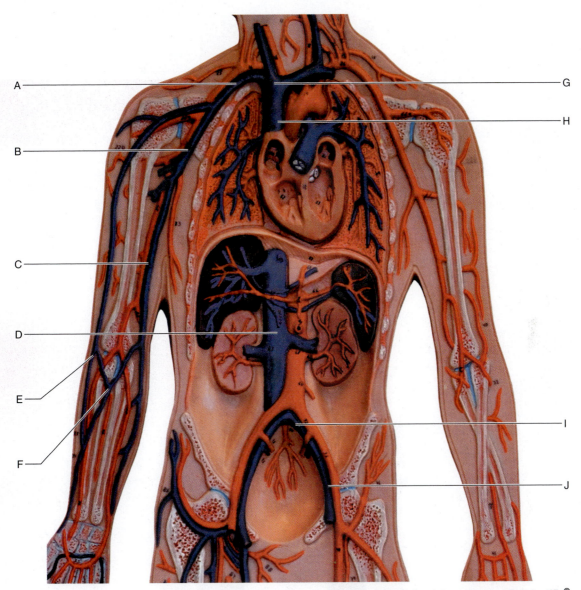

LT-G30: Human circulatory system, 3B Scientific®

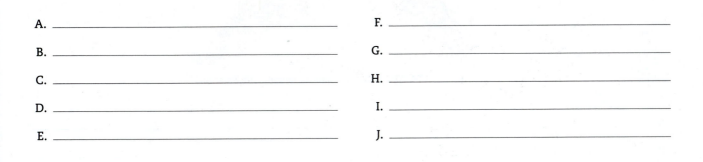

A. _____   F. _____

B. _____   G. _____

C. _____   H. _____

D. _____   I. _____

E. _____   J. _____

Exercise 6.25 Veins of the Pelvis and Lower Limb

GO TO ANATOMICAL MODELS > CARDIOVASCULAR SYSTEM > VEINS > SELF REVIEW > IMAGE 18

- *Mouse over the image to locate and label the vessels indicated below. Click on the vessels to hear their pronunciations.*

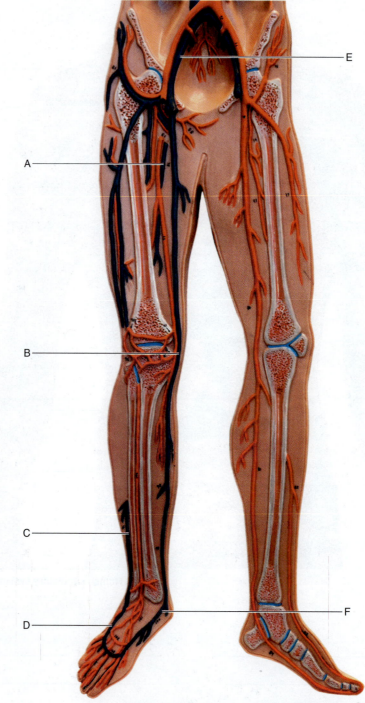

LT-G30: Human circulatory system, 3B Scientific®

A. _____   D. _____

B. _____   E. _____

C. _____   F. _____

QUIZ) Anatomical Models

Answer the following questions using information from PAL 3.1 as well as other course materials including your textbook, lecture, and lab notes.

I. Check Your Understanding

1. The common iliac vein is formed by the union of which two vessels?

_____

_____

2. Which vessel(s) transport blood away from the confluence of the sinuses in the dura mater?

_____

_____

3. Which two veins join to form the brachiocephalic vein?

_____

_____

4. The great saphenous vein drains into which vessel?

_____

BEYOND PAL **5.** Which veins transport oxygenated blood?

_____

_____

6. Into which vessel does the azygos vein drain?

_____

II. Apply What You Learned

1. The great saphenous vein is one vessel that may be used in a heart bypass surgery. Heart bypass surgery is performed when there is a partial to complete blockage of an artery supplying blood to the heart wall.

a. Why would this vein be used as a bypass for arteries?

_____

_____

_____

_____

_____

b. What feature of a vein requires a specific orientation of the vessel when it is used for a bypass?

_____

_____

2. After a right knee replacement surgery, a 67-year-old male suffers from a left pulmonary artery embolism, which is an obstruction of the pulmonary artery by a blood clot. Emergency surgery is successfully performed to remove the clot. Describe the path the clot had taken, starting at the right femoral vein and ending at the left pulmonary artery.

_____

_____

_____

_____

_____

_____

_____

TISSUES OF THE CARDIOVASCULAR SYSTEM

SELF REVIEW | Histology

Exercise 6.26 **Blood Smear 400x**

GO TO › HISTOLOGY > CARDIOVASCULAR SYSTEM > SELF REVIEW > IMAGE 1

• *Mouse over the image to locate and label the structures indicated below. Click on the structures to hear their pronunciations.*

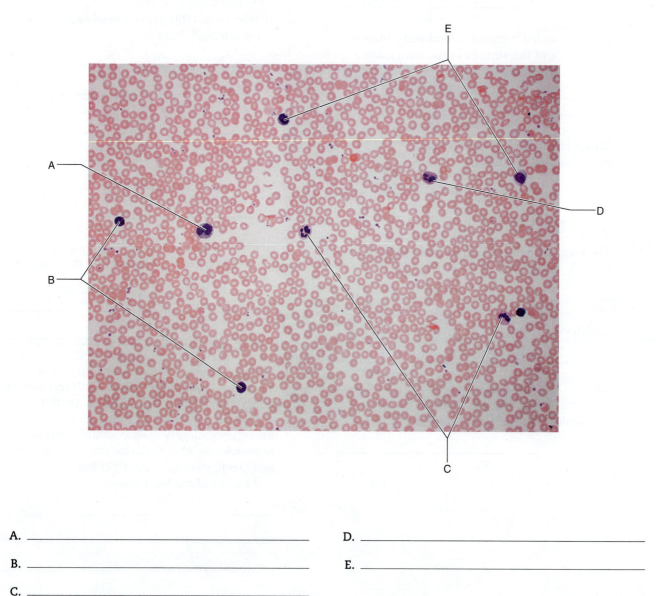

A. _____

B. _____

C. _____

D. _____

E. _____

Exercise 6.27   **Artery and Vein 40x and 100x**

 HISTOLOGY > CARDIOVASCULAR SYSTEM > SELF REVIEW > IMAGES 8 AND 9

- *Mouse over the images to locate and label the structures indicated below. Click on the structures to hear their pronunciations.*

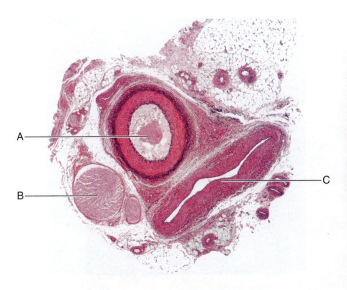

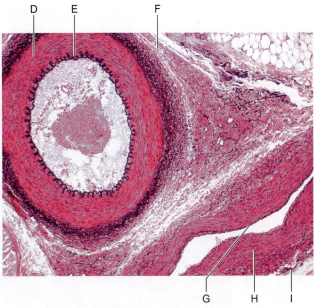

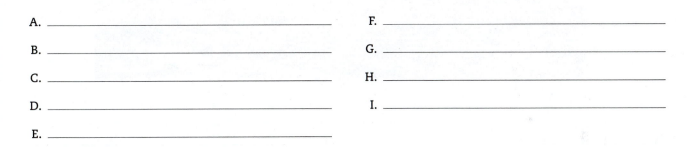

A. _____

B. _____

C. _____

D. _____

E. _____

F. _____

G. _____

H. _____

I. _____

Exercise 6.28 **Cardiac Muscle Tissue 400x**

GO TO ▷ HISTOLOGY > CARDIOVASCULAR SYSTEM > SELF REVIEW > IMAGE 19

• *Mouse over the image to locate and label the structures indicated below. Click on the structures to hear their pronunciations.*

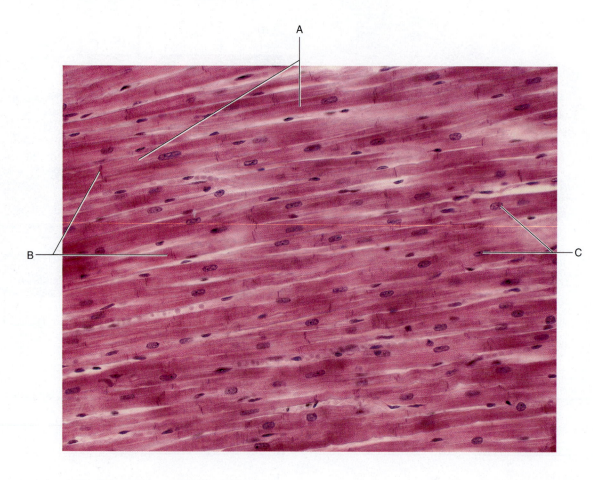

A. _____   C. _____

B. _____

QUIZ ⟩ Histology

Answer the following questions using information from PAL 3.1 as well as other course materials including your textbook, lecture, and lab notes.

I. Check Your Understanding

1. Which type of blood vessel contains valves?

   _____

   _____

2. Valves are extensions of which layer of a blood vessel?

   _____

   _____

3. In a cross section of an artery and vein from the same region (subclavian artery and vein, for example), which type of vessel will have a larger diameter?

   _____

   _____

4. Which structure connects adjacent cardiac muscle cells?

   _____

   _____

5. Which type of blood cell predominates in a blood smear?

   _____

   _____

BEYOND PAL 6. Which layer of a blood vessel wall contains smooth muscle?

   _____

   _____

7. Which layer of a blood vessel contains the endothelium?

   _____

   _____

8. Which type of white blood cell is the most abundant?

   _____

9. Which type of blood cell consists of both B cells and T cells?

   _____

   _____

II. Apply What You Learned

Your friend Sam tells you that, as a way to make a little extra money, he is planning to donate plasma twice a week. He has never donated blood or plasma, but he knows you are taking a course in human anatomy. How would you answer Sam's following questions?

1. Why is blood collected from the vein in my arm instead of taking blood from an artery?

   _____

   _____

   _____

   _____

   _____

2. If arteries contain blood under higher pressure, wouldn't it be easier to extract blood from an artery?

   _____

   _____

   _____

   _____

3. What is the difference between donating blood and donating plasma?

   _____

   _____

   _____

   _____

   _____

   _____

   _____

   _____

LAB PRACTICAL Cardiovascular System

1. Identify the vessel.

   _____

2. Identify the vessel.

   _____

3. Identify the vessel.

   _____

4. Identify the structure.

   _____

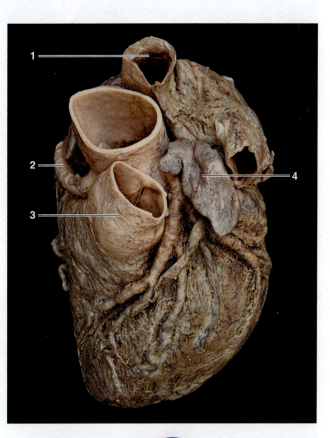

5. Identify the vessel.

   _____

6. Identify the vessel.

   _____

7. Identify the vessel.

   _____

8. Into which chamber is this vessel emptying?

   _____

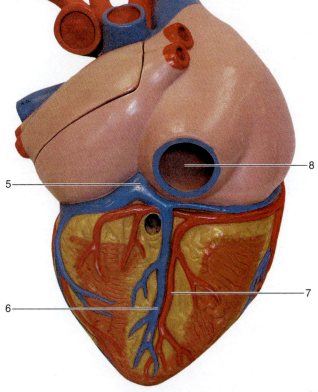

LT-G12: Giant heart, 4-part, 3B Scientific®

LAB PRACTICAL *continues*

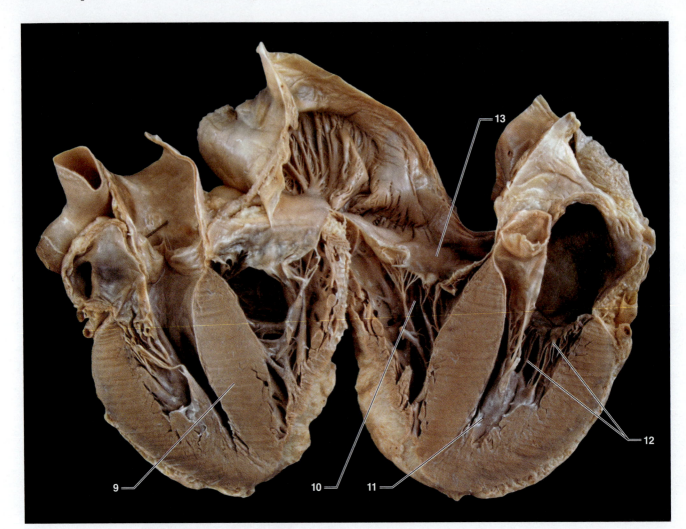

9. Identify the structure.

_____

10. Identify the chamber.

_____

11. Identify the structure.

_____

12. Identify the structures.

_____

13. Identify the structure.

_____

14. Identify the vessel.

_____

15. Which organ is supplied by this vessel?

_____

16. Identify the vessel.

_____

17. Identify the vessel.

_____

18. Identify the vessel.

_____

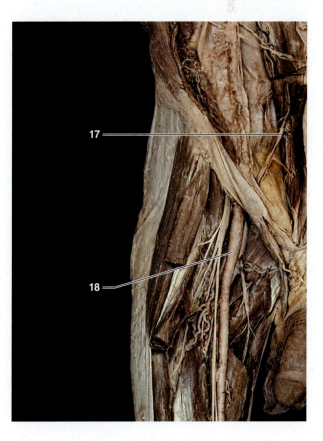

LAB PRACTICAL *continues*

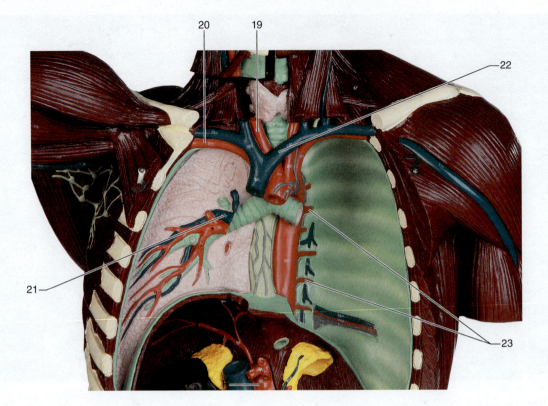

19. Identify the vessel.

_____

20. Identify the vessel.

_____

21. Identify the vessel.

_____

22. Identify the vessel.

_____

23. Identify the vessels.

_____

24. Identify the vessel.

_____

25. Identify the vessel.

_____

26. Identify the vessel.

_____

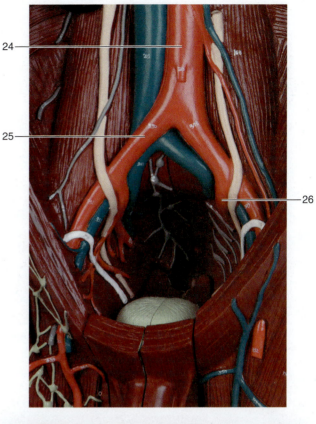

27. This vessel drains into which larger vein?

_____

28. Identify the vessel.

_____

29. Identify the vessel.

_____

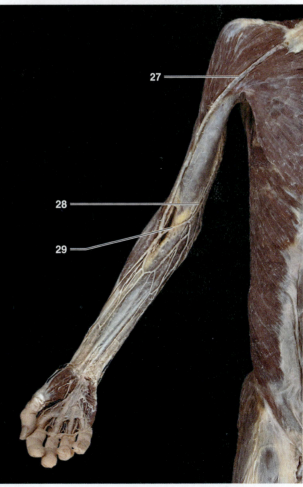

LAB PRACTICAL _continues_

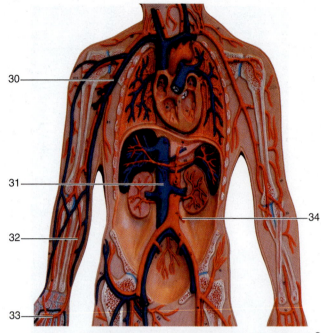

LT-G30: Human circulatory system, 3B Scientific®

30. Identify the vessel.

_____

31. Identify the vessel.

_____

32. Identify the vessel.

_____

33. Identify the vessel.

_____

34. Identify the vessel.

_____

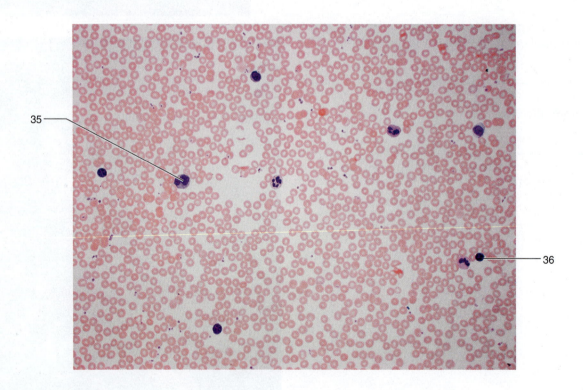

35. Identify the cell type.

_____

36. Identify the cell type.

_____

37. What cell type predominates in this layer?

_____

38. Identify the layer.

_____

39. Identify the type of blood vessel.

_____

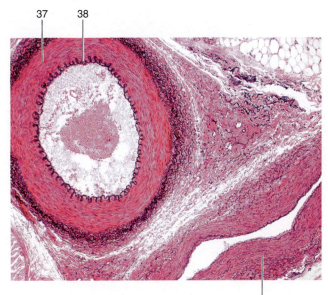

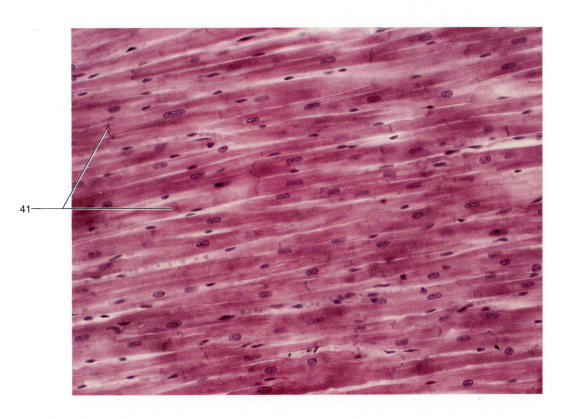

40. Identify the cell type in the image above.

_____

41. Identify the structures.

_____

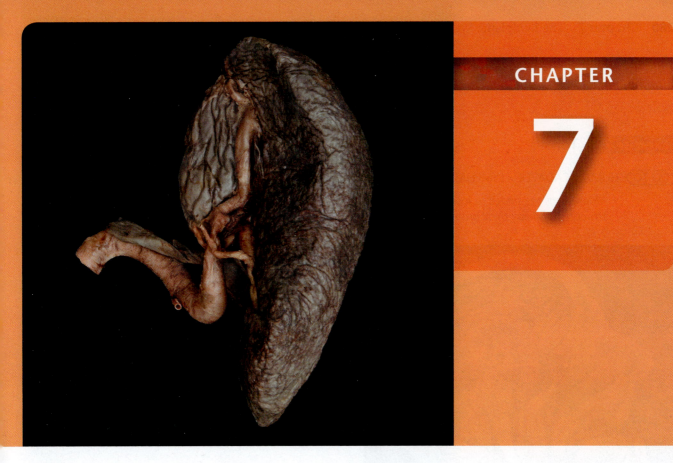

The Lymphatic System

STUDENT OBJECTIVES

GROSS ANATOMY OF THE LYMPHATIC SYSTEM

1. Identify the major structures and organs of the lymphatic system.
2. Describe the organization and distribution of lymphatic vessels.
3. Explain how lymph is transported from tissues back to the blood.
4. Understand the distribution of lymph nodes throughout the body.

TISSUES OF THE LYMPHATIC SYSTEM

5. Describe the structures and organization of a lymph node.
6. Describe the structures and organization of the spleen.
7. Describe the structures and organization of the thymus gland.
8. Describe the structures and organization of the palatine tonsil.
9. Describe the location and function of Peyer's patches.

GROSS ANATOMY OF THE LYMPHATIC SYSTEM

SELF REVIEW | Human Cadaver

Exercise 7.1 Tonsils

GO TO > HUMAN CADAVER > LYMPHATIC SYSTEM > SELF REVIEW > IMAGE 1

- *Mouse over the image to locate and label the structures indicated below. Click on the structures to hear their pronunciations.*

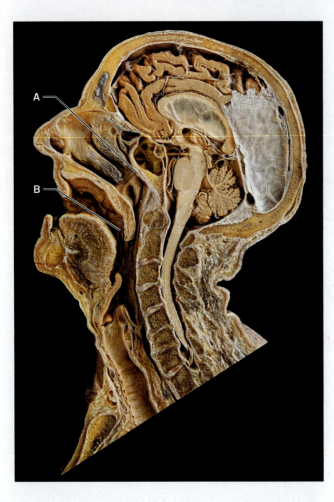

A. _____

B. _____

Exercise 7.2   Lymphatic Structures

GO TO ⟩ HUMAN CADAVER > LYMPHATIC SYSTEM > SELF REVIEW > IMAGES 4 AND 5

- *Mouse over the images to locate and label the structures indicated below. Click on the structures to hear their pronunciations.*

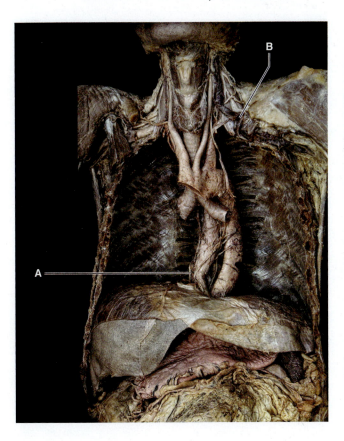
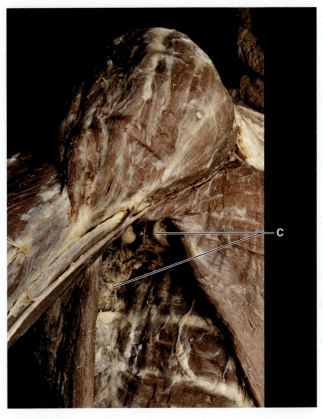

A. _____

B. _____

C. _____

Exercise 7.3 Spleen and Lymphatic Structures

GO TO HUMAN CADAVER > LYMPHATIC SYSTEM > SELF REVIEW > IMAGES 8 AND 9

- *Mouse over the images to locate and label the structures indicated below. Click on the structures to hear their pronunciations.*

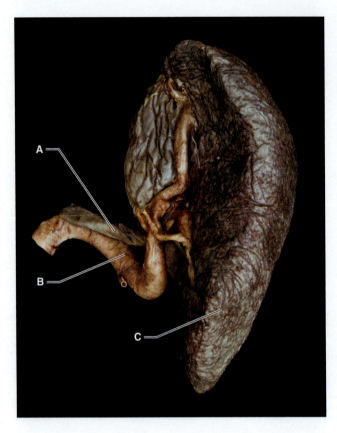

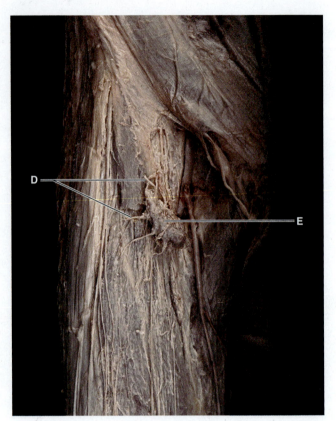

A. _____   D. _____

B. _____   E. _____

C. _____

QUIZ Human Cadaver

Answer the following questions using information from PAL 3.1 as well as other course materials including your textbook, lecture, and lab notes.

I. Check Your Understanding

1. The right upper limb and right side of the head and neck drain into which lymphatic duct?

_____

_____

2. Into which vein does the thoracic duct empty lymph?

_____

_____

3. The cisterna chyli is an enlarged sac that empties into the thoracic duct and can be found near which region of the aorta?

_____

_____

4. Which lymph nodes are located in the neck?

_____

_____

BEYOND PAL

5. Place the following structures in the order in which lymph would pass through them: **lymph duct**, **lymphatic collecting vessel**, **lymph trunk**, **lymph capillary.**

   1. _____

   2. _____

   3. _____

   4. _____

6. What fluid is transported within lymphatic vessels?

   _____

7. The lymphatic system plays an important role in being the primary transport mechanism for which of the following absorbed nutrients moving from the intestine to the blood?

   a. carbohydrates

   b. lipids (fats)

   c. vitamins

   d. proteins

II. Apply What You Learned

Lymphatic filariasis is a condition caused by a parasitic nematode worm that inhabits lymphatic vessels and lymph nodes. Larvae of the parasitic nematode are transmitted when an infected mosquito bites a human.

1. How are the larvae able to gain access to the lymphatic vessels?

   _____

   _____

   _____

   _____

   _____

   _____

   _____

   _____

2. The adult stage of the parasite is a nematode worm. Why might the presence of a worm in the lymphatic vessels be a problem?

   _____

   _____

   _____

   _____

   _____

   _____

3. If a lymphatic vessel is blocked, what would be the result?

   _____

   _____

   _____

   _____

   _____

   _____

SELF REVIEW | Anatomical Models

Exercise 7.4 **Tonsils**

GO TO ANATOMICAL MODELS > LYMPHATIC SYSTEM > SELF REVIEW > IMAGE 1

- *Mouse over the image to locate and label the structures indicated below. Click on the structures to hear their pronunciations.*

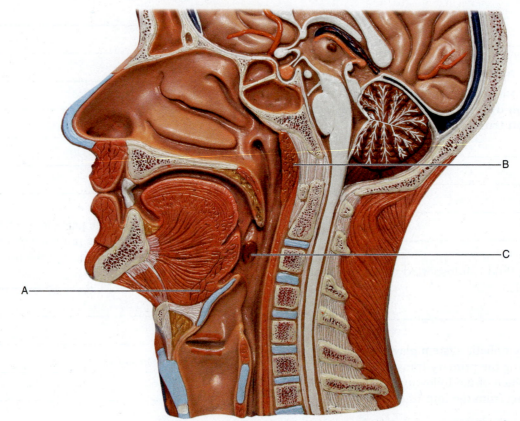

LT-C24: Half head with musculature, 3B Scientific®

A. _____   C. _____

B. _____

Exercise 7.5   **Lymph Drainage of the Upper Body**

GO TO ANATOMICAL MODELS > LYMPHATIC SYSTEM > SELF REVIEW > IMAGE 2

- *Mouse over the image to locate and label the structures indicated below. Click on the structures to hear their pronunciations.*

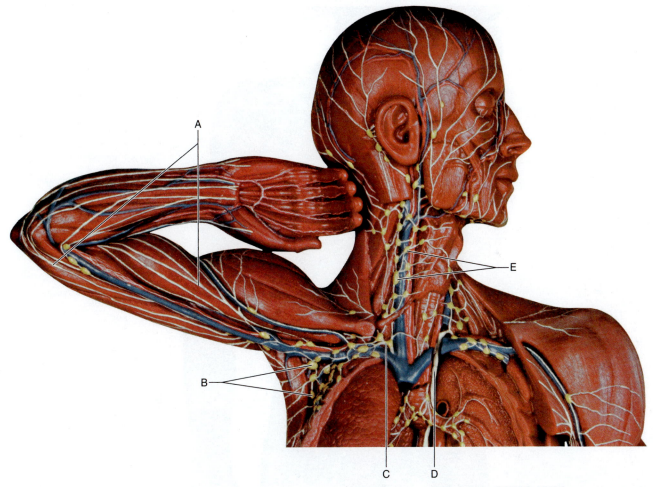

A. _____   D. _____

B. _____   E. _____

C. _____

Exercise 7.6 **Lymph Drainage of the Thoracic and Abdominal Cavities**

GO TO ANATOMICAL MODELS > LYMPHATIC SYSTEM > SELF REVIEW > IMAGE 3

• *Mouse over the image to locate and label the structures indicated below. Click on the structures to hear their pronunciations.*

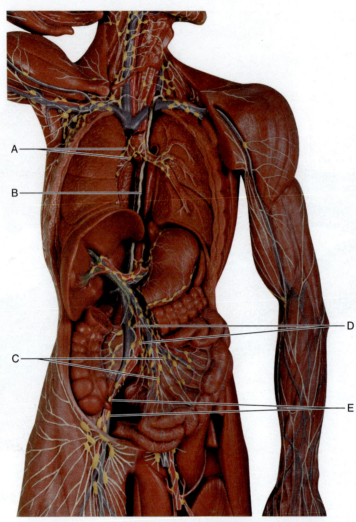

Copyright by SOMSO, 2010, www.somso.com

A. _____   D. _____

B. _____   E. _____

C. _____

Exercise 7.7 Lymph Drainage of the Lower Limb

GO TO ANATOMICAL MODELS > LYMPHATIC SYSTEM > SELF REVIEW > IMAGE 8

- *Mouse over the image to locate and label the structures indicated below. Click on the structures to hear their pronunciations.*

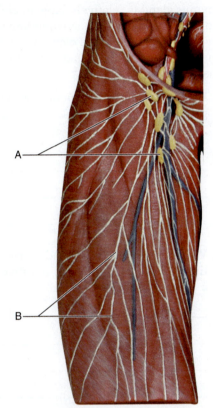

Copyright by SOMSO, 2010, www.somso.com

A. _____

B. _____

QUIZ | Anatomical Models

Answer the following questions using information from PAL 3.1 as well as other course materials including your textbook, lecture, and lab notes.

I. Check Your Understanding

1. Ultimately, lymph is added to what other type of fluid found in the body?

   _____

2. Which tonsil is located on the posterior surface of the tongue?

   _____

3. Which lymphoid organ has a cortex containing lymphoid follicles and a medulla containing medullary cords?

   _____

BEYOND PAL
4. Which structures of the lymphatic system capture interstitial fluid that has accumulated within the tissues of the body?

   _____

   _____

5. Both lymphatic collecting vessels and veins transport fluid under low-pressure conditions. What structures do they share in common to keep fluid moving in one direction?

   _____

   _____

6. A specific type of lymph capillary is found underlying the epithelium of the intestine to allow for the absorption of lipids. What is this special lymph capillary called?

   _____

   _____

7. Which lymphatic duct is responsible for collecting and draining the majority of the lymph back into the blood?

   _____

   _____

II. Apply What You Learned

The thymus gland is relatively large in children and easy to distinguish from the surrounding tissue. However, it slowly atrophies over an individual's lifetime, and by middle age is difficult to distinguish as a distinct organ.

1. What is the primary function of the thymus gland?

   _____

   _____

   _____

   _____

2. Why does the thymus gland "shrink" over time, and why doesn't this negatively impact a person's immune system?

   _____

   _____

   _____

   _____

   _____

   _____

   _____

   _____

3. Vertebrates have developed an adaptive immune response in addition to the more primitive innate immune response. What is the adaptive immune response, and in which of these two divisions would you expect the thymus to play a large role?

   _____

   _____

   _____

   _____

   _____

   _____

   _____

TISSUES OF THE LYMPHATIC SYSTEM

SELF REVIEW | Histology

Exercise 7.8 Lymph Node 40x

GO TO > HISTOLOGY > LYMPHATIC SYSTEM > SELF REVIEW > IMAGE 1

- *Mouse over the image to locate and label the structures indicated below. Click on the structures to hear their pronunciations.*

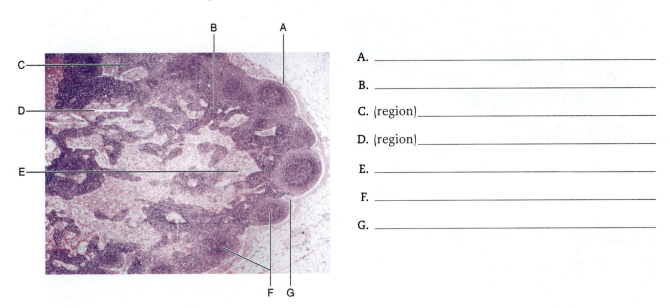

A. _____

B. _____

C. (region)_____

D. (region)_____

E. _____

F. _____

G. _____

Exercise 7.9 Spleen 40x

GO TO > HISTOLOGY > LYMPHATIC SYSTEM > SELF REVIEW > IMAGE 8

- *Mouse over the image to locate and label the structures indicated below. Click on the structures to hear their pronunciations.*

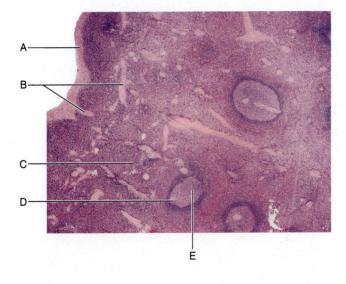

A. _____

B. _____

C. _____

D. _____

E. _____

Exercise 7.10   Thymus Gland 40x

GO TO   HISTOLOGY > LYMPHATIC SYSTEM > SELF REVIEW > IMAGE 12

- *Mouse over the image to locate and label the structures indicated below. Click on the structures to hear their pronunciations.*

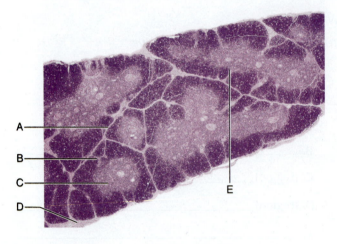

A. _____

B. (region)_____

C. (region)_____

D. _____

E. _____

Exercise 7.11   Palatine Tonsil 40x

GO TO   HISTOLOGY > LYMPHATIC SYSTEM > SELF REVIEW > IMAGE 15

- *Mouse over the image to locate and label the structures indicated below. Click on the structures to hear their pronunciations.*

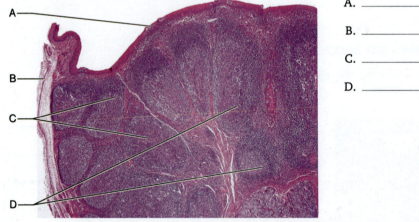

A. _____

B. _____

C. _____

D. _____

Exercise 7.12 **Peyer's Patches in Ileum 40x**

<u>GO TO</u> HISTOLOGY > LYMPHATIC SYSTEM > SELF REVIEW > IMAGE 18

- *Mouse over the image to locate and label the structures indicated below. Click on the structures to hear their pronunciations.*

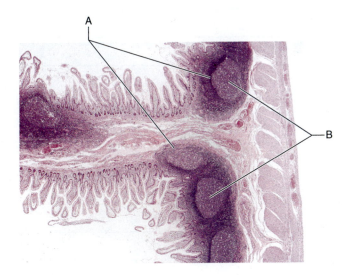

A. _____

B. _____

QUIZ | Histology

Answer the following questions using information from PAL 3.1 as well as other course materials including your textbook, lecture, and lab notes.

I. Check Your Understanding

1. Which lymph nodes are located in the armpit?

_____

_____

2. Which tonsil is located in the nasopharynx?

_____

3. For each of the following organs, list all the structures/regions that are part of that organ. Choose from: **cortex**, **medulla**, **capsule**, **Peyer's patch**, **red pulp**, and **white pulp**. (Note that structures/regions can be used multiple times.)

spleen _____

_____

lymph node _____

_____

ileum _____

_____

thymus gland _____

_____

tonsil _____

_____

BEYOND
PAL

4. B cells are primarily found in which layer of a lymph node?

_____

5. In which structure within a lymph node do B cells proliferate and differentiate during an immune response?

_____

6. Which organ of the lymphatic system is critical in the maturation and differentiation of T cells, and therefore plays a critical role in the adaptive immune response?

_____

7. Which organ of the lymphatic system plays a critical role in filtering the blood of pathogens as well as removing old red blood cells?

_____

8. Which type of tissue, found in the organ from question 7, contains lymphoid follicles seeded with B lymphocytes for filtering the blood?

_____

II. Apply What You Learned

Hodgkin's lymphoma is a type of cancer that originates in lymph nodes, and can spread to other lymph tissues and organs of the body.

1. As a lymphoma, what type of blood cell is cancerous?

_____

2. Why is it that cancer found in lymph nodes can spread so easily to other nodes and organs of the body?

_____

_____

_____

_____

3. A swollen lymph node is typically a sign that your body is fighting off an infection. Why do lymph nodes swell when a pathogen is present?

_____

_____

_____

_____

_____

_____

4. A persistent swollen lymph node is a symptom of Hodgkin's lymphoma. In this situation, what is causing the node to swell?

_____

_____

_____

_____

_____

LAB PRACTICAL Lymphatic System

1. Identify the structure.

   _____

2. Identify the structure.

   _____

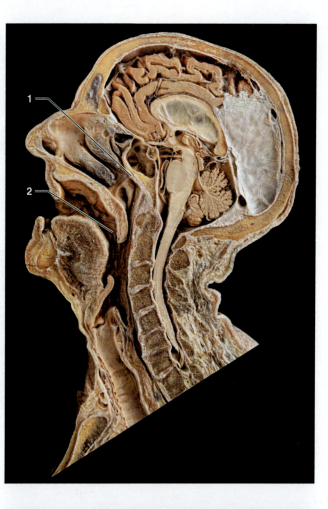

3. Identify the structure.

   _____

4. Identify the organ.

   _____

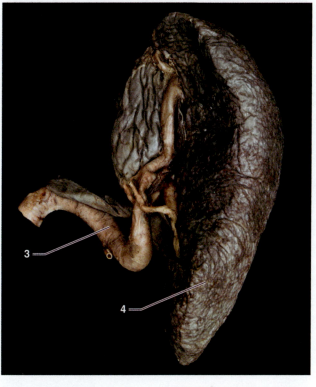

LAB PRACTICAL *continues*

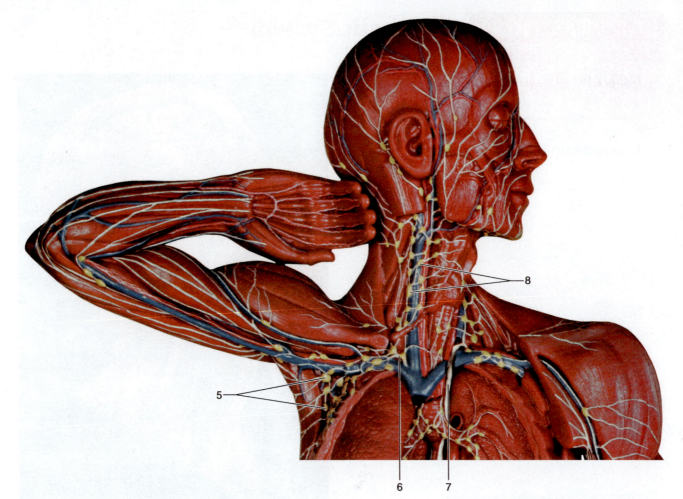

5. Identify the structures.

_____

6. Identify the structure.

_____

7. Identify the structure.

_____

8. Identify the structures.

_____

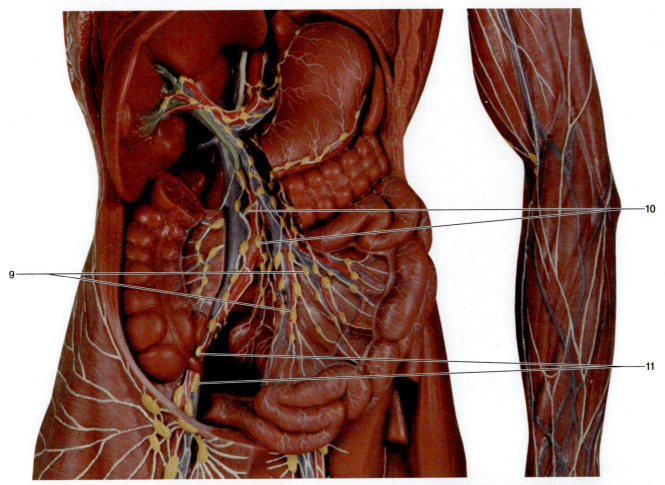

9. Identify the structures.

_____

10. Identify the structures.

_____

11. Identify the structures.

_____

LAB PRACTICAL *continues*

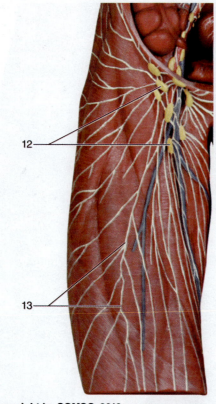

12. Identify the structures.

_____

13. Identify the structures.

_____

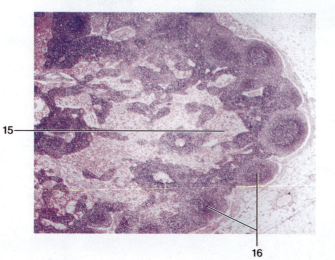

14. The image at left is a section through what type of lymphatic structure?

_____

15. Identify the space.

_____

16. Identify the structures.

_____

17. Identify the structures.

_____

18. Identify the type of tissue.

_____

19. Identify the type of tissue.

_____

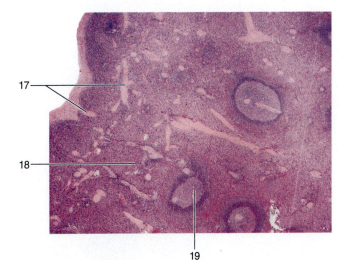

20. Identify the region.

_____

21. Identify the region.

_____

22. Identify the structures.

_____

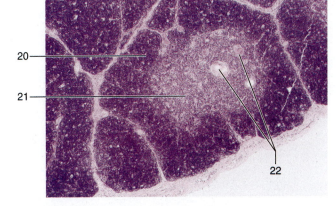

23. Identify the type of epithelium.

_____

24. Identify the structures.

_____

25. Identify the structures (within #24).

_____

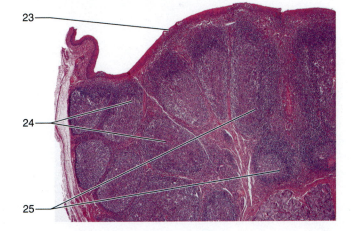

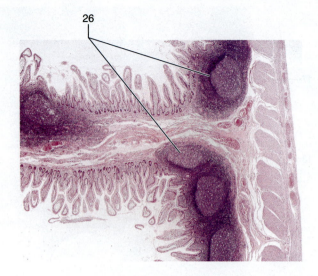

26

26. Identify the structures.

_____

27. The image at left is a section through which specific region of the small intestine?

_____

The Respiratory System

STUDENT OBJECTIVES

GROSS ANATOMY OF THE RESPIRATORY SYSTEM

1. Identify the respiratory passageways in order, from the nose to the alveoli.
2. Distinguish the structures of the conducting zone from those of the respiratory zone.
3. Identify the structures of the nasal cavity.
4. Identify the tonsils.
5. Identify the structures that make up the larynx.
6. Describe the gross structure of the lungs.
7. Identify the muscles involved in respiration.

TISSUES OF THE RESPIRATORY SYSTEM

8. Understand the basic organization of the trachea wall.
9. Identify the mucus-secreting cells.
10. Describe the structure of a lung alveolus and respiratory membrane.
11. Identify the surfactant-secreting cells, and understand the role of surfactant in respiration.

GROSS ANATOMY OF THE RESPIRATORY SYSTEM

SELF REVIEW | Human Cadaver

Exercise 8.1   **Respiratory Structures of the Head and Neck**

GO TO ⟩ HUMAN CADAVER > RESPIRATORY SYSTEM > SELF REVIEW > IMAGES 1 AND 2

- *Mouse over the images to locate and label the structures below. Click on the structure to hear the pronunciation.*

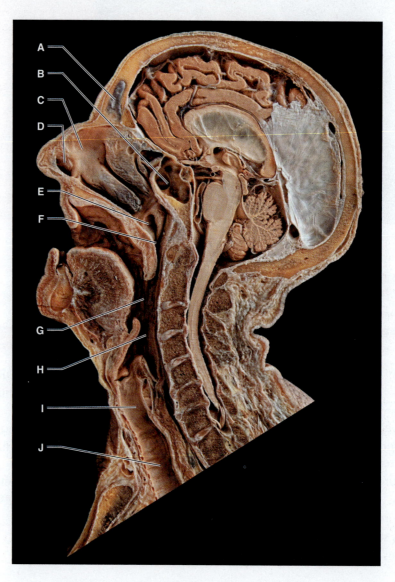

A. _____   F. (space) _____

B. _____   G. (space) _____

C. _____   H. (space) _____

D. _____   I. _____

E. _____   J. _____

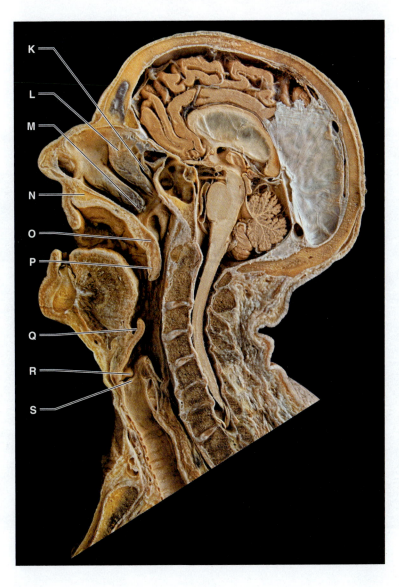

K. _____

L. _____

M. _____

N. _____

O. _____

P. _____

Q. _____

R. _____

S. _____

Exercise 8.2 Larynx

GO TO〉 HUMAN CADAVER > RESPIRATORY SYSTEM > SELF REVIEW > IMAGES 4 AND 5

- *Mouse over the images to locate and label the structures below. Click on the structure to hear the pronunciation.*

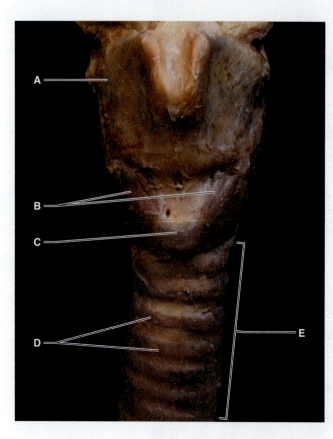 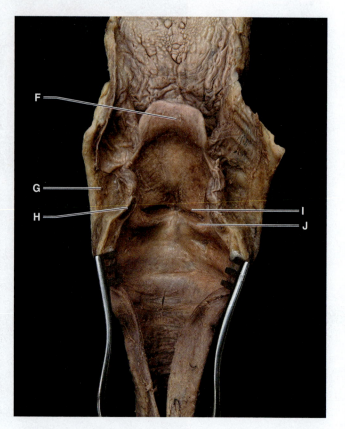

A. _____   F. _____

B. _____   G. _____

C. _____   H. _____

D. _____   I. _____

E. _____   J. _____

Exercise 8.3   Lungs, Anterior View

GO TO HUMAN CADAVER > RESPIRATORY SYSTEM > SELF REVIEW > IMAGE 7

- *Mouse over the image to locate and label the structures below. Click on the structure to hear the pronunciation.*

A. _____   F. _____

B. _____   G. _____

C. _____   H. _____

D. _____   I. _____

E. _____   J. _____

Exercise 8.4 Lungs in the Thoracic Cavity

GO TO HUMAN CADAVER > RESPIRATORY SYSTEM > SELF REVIEW > IMAGE 11

- *Mouse over the image to locate and label the structures below. Click on the structure to hear the pronunciation.*

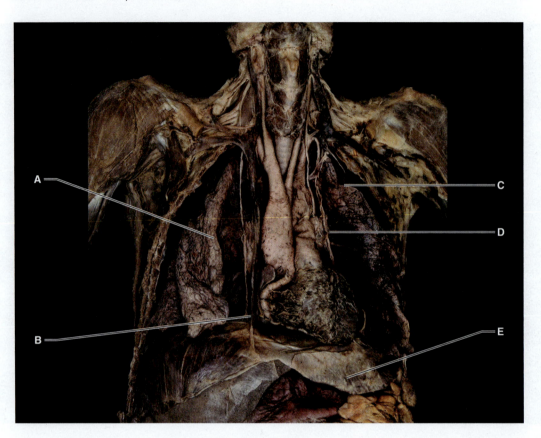

A. _____   D. _____

B. _____   E. _____

C. _____

Exercise 8.5   **Respiratory Structures and Thoracic Cavity (Lungs Removed)**

GO TO   HUMAN CADAVER > RESPIRATORY SYSTEM > SELF REVIEW > IMAGE 13

- *Mouse over the image to locate and label the structures below. Click on the structure to hear the pronunciation.*

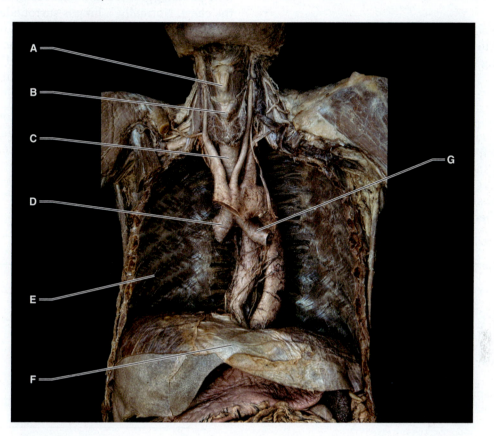

A. _____   E. _____

B. _____   F. _____

C. _____   G. _____

D. _____

QUIZ | Human Cadaver

Answer the following questions using information from PAL 3.1 as well as other course materials including your textbook, lecture, and lab notes.

I. Check Your Understanding

1. How many lobes does the right lung have? The left lung? Name them.

right _____

left _____

2. What are the branches of the right main (primary) bronchus?

_____

_____

3. Which cartilage of the larynx has a prominence known as the "Adam's apple"?

_____

_____

4. Which structure covers the opening to the larynx to prevent food and/or liquid from entering the glottis when swallowing?

_____

_____

BEYOND PAL

5. For each respiratory system structure below, indicate which type of epithelium lines it. Choose from **simple squamous**, **pseudostratified ciliated columnar**, or **stratified squamous**.

nasal cavity   _____

nasopharynx   _____

oropharynx   _____

laryngopharynx   _____

trachea   _____

alveolus   _____

6. The respiratory diaphragm is innervated by which nerve?

_____

_____

7. Which group of muscles assists the diaphragm during respiration to elevate the ribs and expand the thoracic cavity?

_____

_____

II. Apply What You Learned

A pneumothorax is a potentially life-threatening condition. It can occur spontaneously without any apparent cause, it can be the result of an underlying lung pathology, or it can be the result of a trauma to the chest. Normally, a lung is surrounded by a thin pleura and serous fluid. The air pressure within the lung is normally higher than that in the pleural space surrounding the lung, so the lung fills the thoracic cavity. In a pneumothorax, however, air or gas fills the pleural cavity around one or both lungs.

1. If there is no natural opening from the lungs into the pleural space surrounding the lungs, how might air get into that space?

_____

_____

_____

_____

_____

2. Why would air occupying the space around the lungs cause a problem?

_____

_____

_____

_____

3. Treatment varies depending on the severity and cause of the pneumothorax. Minor cases may resolve on their own without any treatment. In more serious cases, a chest tube might be inserted to allow for the removal of the trapped air. Where would be the safest place to insert the chest tube?

_____

_____

_____

_____

SELF REVIEW | Anatomical Models

Exercise 8.6 Respiratory Spaces and Regions of the Head and Neck

GO TO > ANATOMICAL MODELS > RESPIRATORY SYSTEM > SELF REVIEW > IMAGE 1

• *Mouse over the image to locate and label the structures below. Click on the structure to hear the pronunciation.*

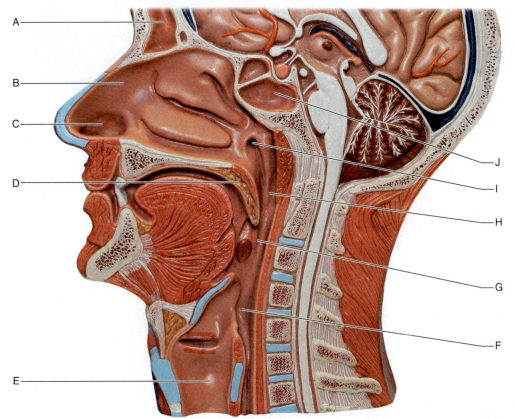

LT-C14: Half head with musculature, 3B Scientific®

A. _____   F. _____

B. _____   G. _____

C. _____   H. _____

D. _____   I. _____

E. _____   J. _____

Exercise 8.7 Respiratory Structures of the Head and Neck

GO TO ANATOMICAL MODELS > RESPIRATORY SYSTEM > SELF REVIEW > IMAGE 2

• *Mouse over the image to locate and label the structures below. Click on the structure to hear the pronunciation.*

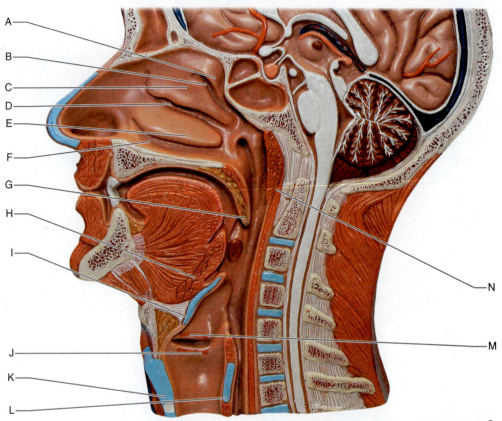

LT-C14: Half head with musculature, 3B Scientific®

A. _____   H. _____

B. (space) _____   I. _____

C. _____   J. _____

D. (space) _____   K. _____

E. _____   L. _____

F. (space) _____   M. _____

G. _____   N. _____

Exercise 8.8 **Larynx, Anterior and Posterior Views**

GO TO ANATOMICAL MODELS > RESPIRATORY SYSTEM > SELF REVIEW > IMAGES 3 AND 4

- *Mouse over the images to locate and label the structures below. Click on the structure to hear the pronunciation.*

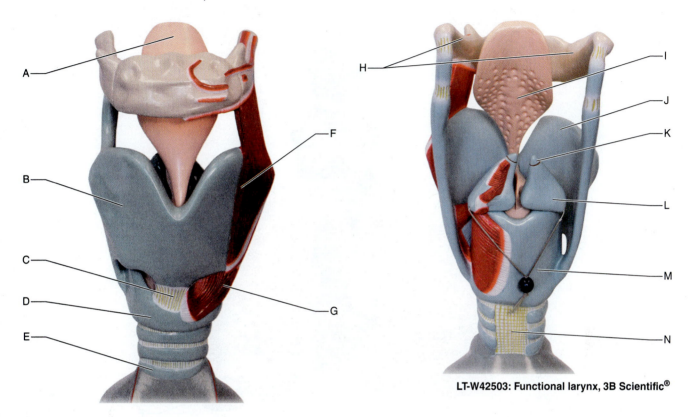

LT-W42503: Functional larynx, 3B Scientific®

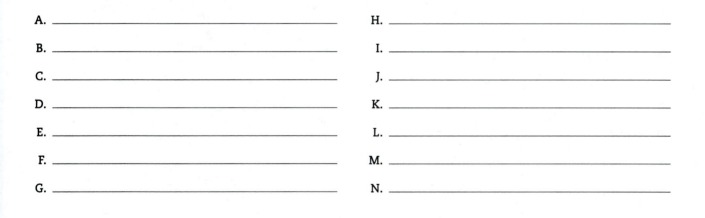

A. _____	H. _____
B. _____	I. _____
C. _____	J. _____
D. _____	K. _____
E. _____	L. _____
F. _____	M. _____
G. _____	N. _____

Exercise 8.9 Lungs

GO TO ANATOMICAL MODELS > RESPIRATORY SYSTEM > SELF REVIEW > IMAGES 7 AND 8

- *Mouse over the images to locate and label the structures below. Click on the structure to hear the pronunciation.*

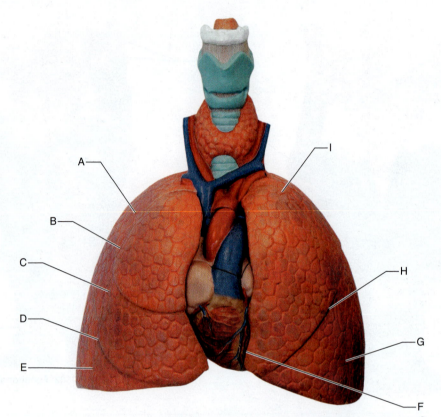

LT-VC243: Lung model with larynx, 5-part, 3B Scientific®

A. _____

B. (space) _____

C. _____

D. (space) _____

E. _____

F. _____

G. _____

H. (space) _____

I. _____

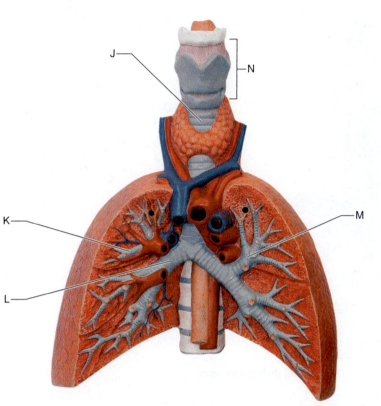

LT-VC243: Lung model with larynx, 5-part, 3B Scientific®

J. _____

K. _____

L. _____

M. _____

N. _____

GO TO ANATOMICAL MODELS > RESPIRATORY SYSTEM > SELF REVIEW > IMAGE 11

> • *Mouse over the image to locate and label the structures below. Click on the structure to hear the pronunciation.*

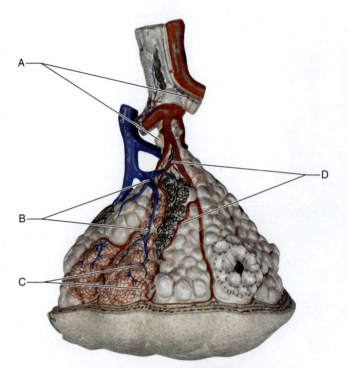

A. _____

B. _____

C. _____

D. _____

Copyright by SOMSO, 2010, www.somso.com

QUIZ Anatomical Models

Answer the following questions using information from PAL 3.1 as well as other course materials including your textbook, lecture, and lab notes.

I. Check Your Understanding

1. Which structure allows air to travel between the pharynx and middle ear?

_____

_____

2. The epiglottis is composed of what type of cartilage?

_____

_____

3. Which lung(s) possess(es) a horizontal fissure?

_____

_____

4. To which cartilage segments of the larynx are the vocal folds attached?

anterior attachment _____

posterior attachment _____

5. What type of epithelium lines the frontal sinus?

_____

6. Indicate whether the following regions belong to the **respiratory zone** or the **conducting zone**.

alveolar sac _____

nasal cavity _____

pharynx _____

trachea _____

larynx _____

respiratory bronchiole _____

main bronchus _____

7. Which two structures are connected via a respiratory membrane? What occurs across this membrane?

_____

_____

_____

_____

II. Apply What You Learned

1. A deviated nasal septum can occur as a result of a hard blow to the nose, or deviation may occur in an infant during childbirth.

a. Where is the nasal septum located?

_____

_____

b. What problems might occur as a result of a deviated septum?

_____

_____

_____

_____

c. In severe cases, a surgery known as septoplasty may be performed to alleviate the condition. What do you think this procedure involves?

_____

_____

_____

_____

2. While hiking with a group of friends, a 21-year-old female chokes on food and a blockage occurs at the glottis. The Valsalva maneuver is performed, but does not clear the food that is blocking her airway. An emergency tracheotomy is successfully performed by a third-year medical student. He makes a thin incision and inserts a tube into the part of the larynx between the cricoid and thyroid cartilages. This procedure saves her life.

What is the name of the thin membrane that was cut through in order to insert the tube?

_____

_____

TISSUES OF THE RESPIRATORY SYSTEM

SELF REVIEW | Histology

Exercise 8.11 Trachea 400x

GO TO ⟩ HISTOLOGY > RESPIRATORY SYSTEM > SELF REVIEW > IMAGE 3

- *Mouse over the image to locate and label the structures below. Click on the structure to hear the pronunciation.*

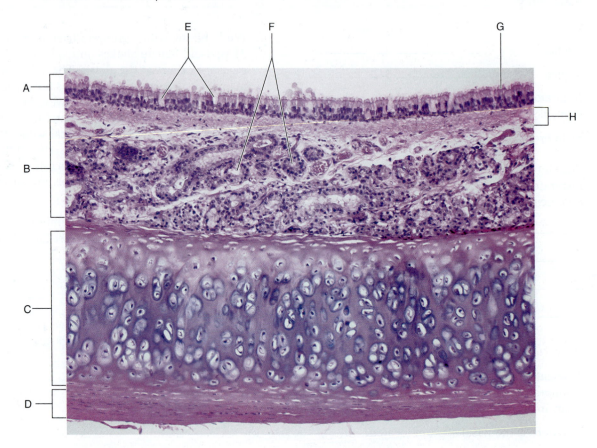

A. _____   E. _____

B. _____   F. _____

C. _____   G. _____

D. _____   H. _____

Exercise 8.12   **Lung 40x and 100x**

GO TO HISTOLOGY > RESPIRATORY SYSTEM > SELF REVIEW > IMAGES 6 AND 7

- *Mouse over the images to locate and label the structures below. Click on the structure to hear the pronunciation.*

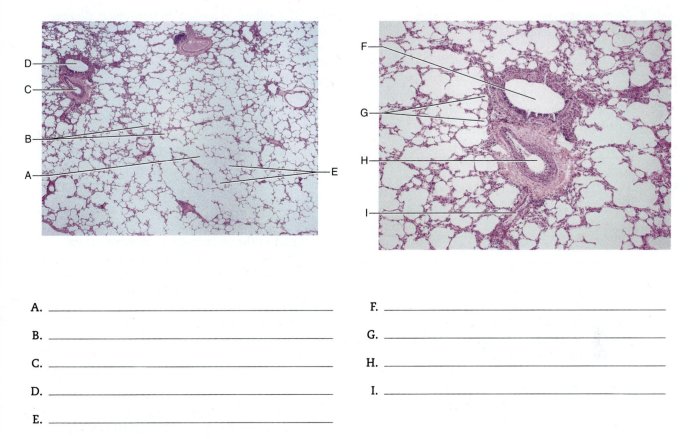

A. _____   F. _____

B. _____   G. _____

C. _____   H. _____

D. _____   I. _____

E. _____

Exercise 8.13   **Alveolar Wall 1000x**

GO TO HISTOLOGY > RESPIRATORY SYSTEM > SELF REVIEW > IMAGE 10

- *Mouse over the image to locate and label the structures below. Click on the structure to hear the pronunciation.*

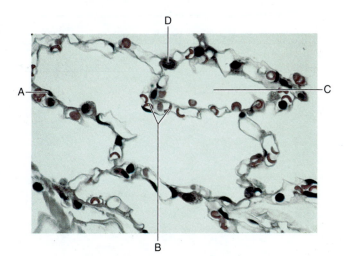

A. _____

B. _____

C. _____

D. _____

QUIZ) Histology

Answer the following questions using information from PAL 3.1 as well as other course materials including your textbook, lecture, and lab notes.

I. Check Your Understanding

1. Seromucous glands are located within which layer of the trachea?

   _____

2. The trachea is kept open by cartilage rings. These rings are composed of what type of cartilage?

   _____

BEYOND PAL

3. Which type of alveolar cell secretes surfactant? What is the purpose of surfactant?

   _____

   _____

   _____

4. Which cells located within the epithelium contribute mucus to the secretions lining most of the structures of the respiratory system?

   _____

   _____

5. What structure is formed by the fusion of simple squamous epithelium of an alveolus and simple squamous epithelium of a capillary via a basement membrane?

   _____

   _____

6. What type of tissue composes the lamina propria?

   _____

7. Do the pulmonary arteries contain oxygenated or deoxygenated blood?

   _____

II. Apply What You Learned

Bronchial asthma is a reaction to inhaled allergens such as pollen, mold, or dust. In individuals who are hypersensitive to these allergens, the body experiences an inflammatory response in the bronchial tubes, and muscle surrounding the larger bronchi constrict.

1. Which type of muscle is found within the bronchi? Is the muscle relaxed or constricted when under sympathetic control?

   _____

   _____

2. Part of the inflammatory response is an increase in mucosal secretions into the bronchi due to the release of histamine from mast cells. Which bronchial cells or structures increase their mucosal secretions in response to histamine?

   _____

   _____

3. Which do you think would be the best initial treatment for chronic bronchial asthma: a bronchodilator to relax the smooth muscle OR an anti-inflammatory such as glucocorticoids? Why?

   _____

   _____

   _____

   _____

   _____

   _____

LAB PRACTICAL Respiratory System

1. Identify the space.

   _____

2. Identify the structure.

   _____

3. Identify the space.

   _____

4. Identify the structure.

   _____

5. Identify the structure.

   _____

6. Identify the space.

   _____

7. Identify the structure.

   _____

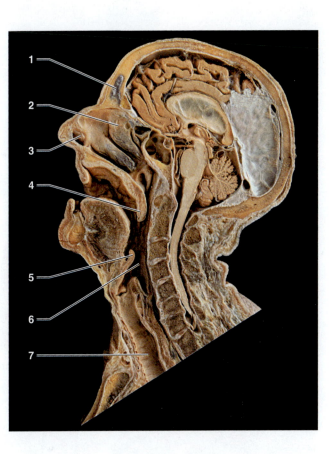

8. Identify the structure.

   _____

9. Identify the structure.

   _____

10. Identify the structure.

    _____

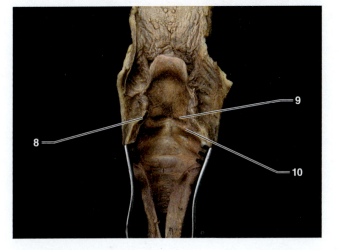

LAB PRACTICAL *continues*

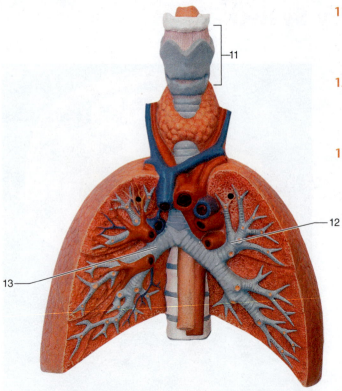

LT-VC243: Lung model with larynx, 5-part, 3B Scientific®

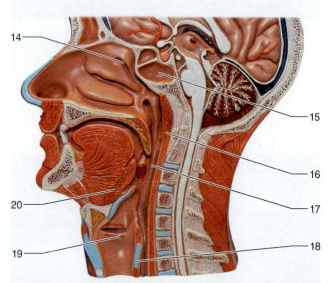

LT-C14: Half head with musculature, 3B Scientific®

11. Identify the structure.

_____

12. Identify the structure.

_____

13. Identify the structure.

_____

14. Identify the space.

_____

15. Identify the space.

_____

16. Identify the structure.

_____

17. Identify the space.

_____

18. Identify the specific cartilage.

_____

19. Identify the structure.

_____

20. Identify the structure.

_____

21. Identify the structure.

_____

22. Identify the structure.

_____

23. Identify the structure.

_____

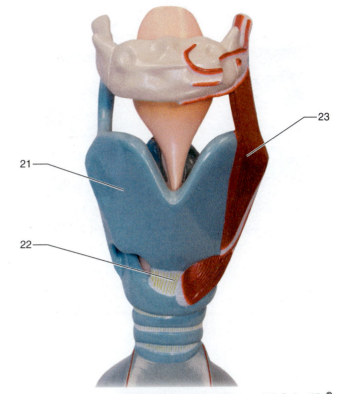

LT-W42503: Functional larynx, 3B Scientific®

24. Identify the region of the lung.

_____

25. Identify the lobe.

_____

26. Identify the space.

_____

27. Identify the space.

_____

28. Identify the lobe.

_____

LAB PRACTICAL _continues_

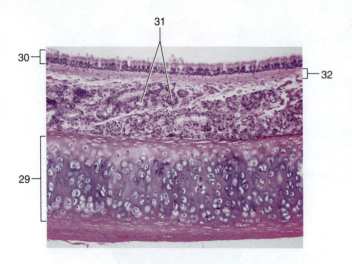

29. Identify the type of cartilage.

_____

30. Identify the type of epithelium.

_____

31. Identify the structures.

_____

32. Identify the layer.

_____

The Digestive System

STUDENT OBJECTIVES

GROSS ANATOMY OF THE DIGESTIVE SYSTEM

1. Identify the organs of the alimentary canal and the accessory digestive organs.
2. Trace the path of food as it passes from the oral cavity through the large intestine.
3. List the major processes that occur during digestion.
4. Name organs involved in propulsion, segmentation, chemical digestion, mechanical digestion, and absorption.
5. Identify each salivary gland.
6. Identify the different types of teeth.
7. Identify the different regions and structures of the stomach, and explain the role that rugae play in the ability of the stomach to expand.
8. Identify the different regions of the small intestine, and explain how they are both similar and distinct from one another.
9. Name the different mesenteries that support the organs within the abdominal cavity.
10. Identify the different ducts that connect the liver, gallbladder, and duodenum and explain the importance of those ducts.
11. Identify the different regions of the pancreas, and understand the role the pancreas has in digestion.

TISSUES OF THE DIGESTIVE SYSTEM

12. Identify and describe the four layers of the wall of the alimentary canal.
13. Describe the arrangement of the smooth muscle layers found within the alimentary canal.
14. Identify the location and function of the major cells and structures that aid in digestion, including goblet cells, parietal cells, chief cells, hepatocytes, serous acini, seromucous glands, duodenal glands, gastric pits, and intestinal crypts.
15. Identify the different epithelial types lining the lumen of organs in the alimentary canal.
16. Describe a liver lobule.
17. Identify the different cell types found within salivary glands.

GROSS ANATOMY OF THE DIGESTIVE SYSTEM

SELF REVIEW | Human Cadaver

Exercise 9.1 **Digestive Structures of the Head and Neck**

GO TO > HUMAN CADAVER > DIGESTIVE SYSTEM > SELF REVIEW > IMAGES 1 AND 3

- *Mouse over the images to locate and label the structures below. Click on the structures to hear their pronunciations.*

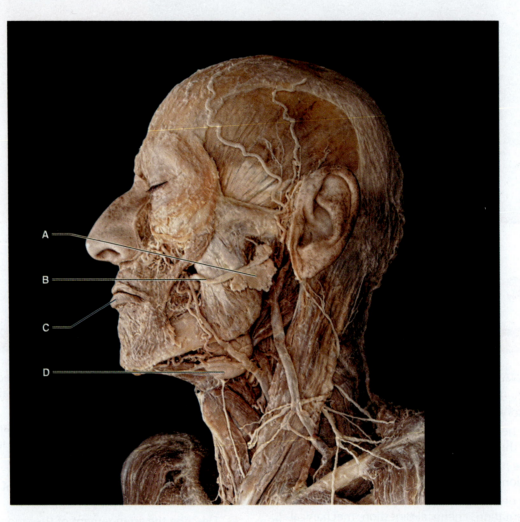

A. _____   C. _____

B. _____   D. _____

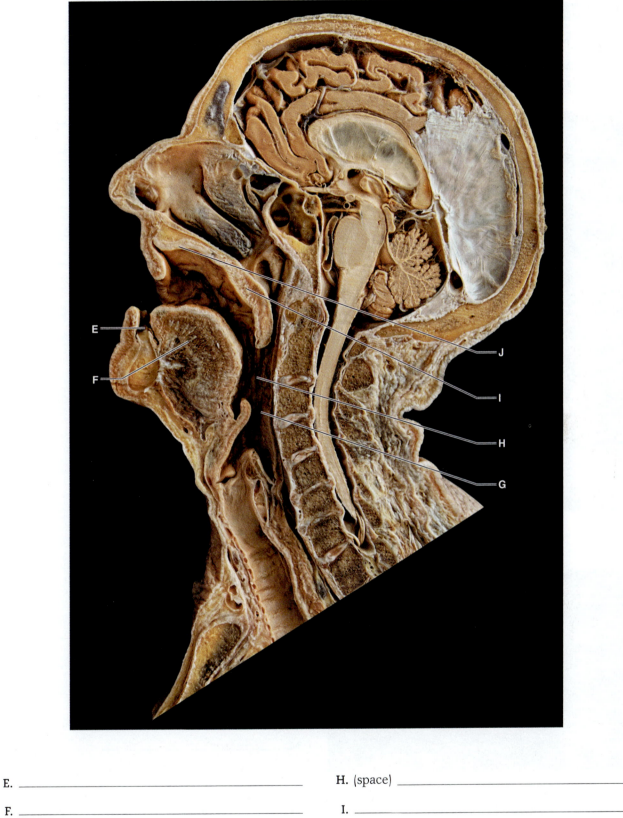

E. _____

F. _____

G. (space) _____

H. (space) _____

I. _____

J. _____

Exercise 9.2 **Digestive Organs of the Abdominal Cavity**

GO TO ⟩ HUMAN CADAVER > DIGESTIVE SYSTEM > SELF REVIEW > IMAGES 9 AND 12

- *Mouse over the images to locate and label the structures below. Click on the structures to hear their pronunciations.*

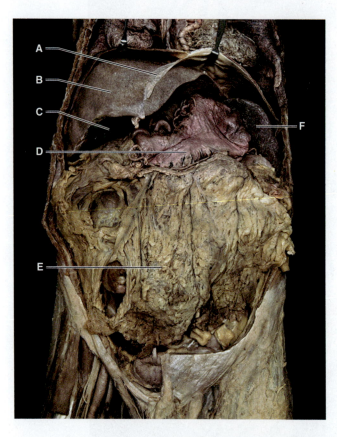

A. _____

B. _____

C. _____

D. _____

E. _____

F. _____

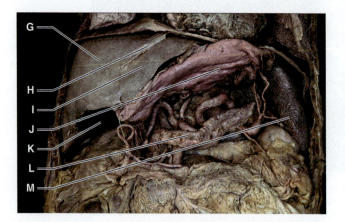

G. _____

H. _____

I. _____

J. _____

K. _____

L. _____

M. _____

Exercise 9.3 Accessory Structures of the Abdominal Cavity

GO TO HUMAN CADAVER > DIGESTIVE SYSTEM > SELF REVIEW > IMAGES 14 AND 15

- *Mouse over the images to locate and label the structures below. Click on the structures to hear their pronunciations.*

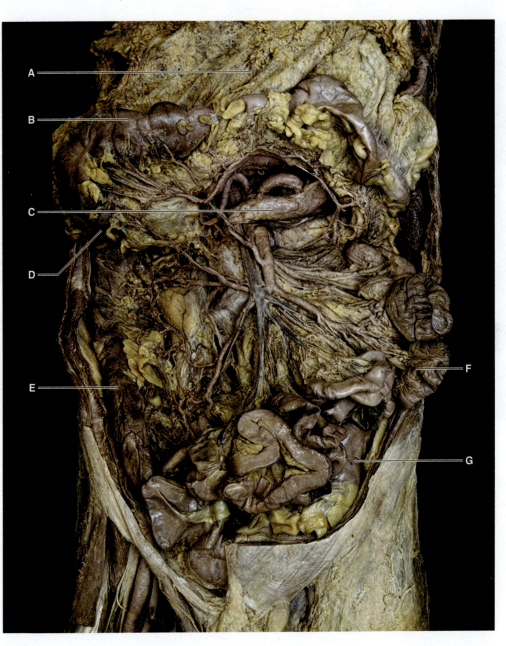

A. _____   E. _____

B. _____   F. _____

C. _____   G. _____

D. _____

Exercise 9.4 Stomach, Liver, Gallbladder, and Pancreas

GO TO > HUMAN CADAVER > DIGESTIVE SYSTEM > SELF REVIEW > IMAGES 16 AND 20

> • *Mouse over the images to locate and label the structures below. Click on the structures to hear their pronunciations.*

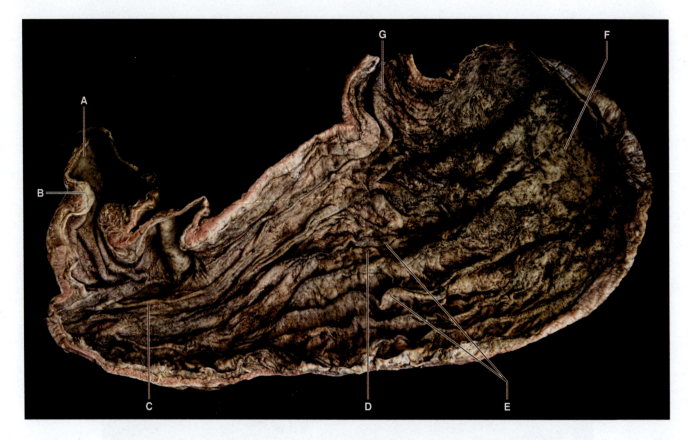

A. _____   E. _____

B. _____   F. _____

C. _____   G. _____

D. _____

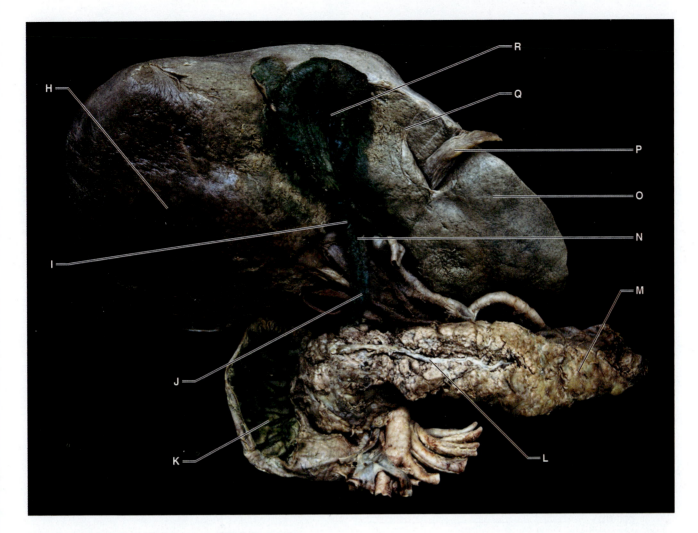

H. _____

I. _____

J. _____

K. _____

L. _____

M. _____

N. _____

O. _____

P. _____

Q. _____

R. _____

QUIZ | Human Cadaver

Answer the following questions using information from PAL 3.1 as well as other course materials including your textbook, lecture, and lab notes.

I. Check Your Understanding

1. Identify the three major parts of the stomach.

_____

_____

_____

2. Into which specific organ does the stomach empty?

_____

3. Which mesentery attaches to the greater curvature of the stomach?

_____

4. For each of the following organs, indicate whether they are part of the **alimentary canal** or whether they are an **accessory digestive organ**, or **neither**.

pancreas _____

stomach _____

cecum _____

spleen _____

liver _____

gallbladder _____

esophagus _____

BEYOND PAL **5.** What is the function of the main pancreatic duct, and where does it empty?

_____

_____

_____

_____

6. What are the teniae coli of the large intestine, and which anatomical feature of the large intestine do they form?

_____

_____

_____

_____

7. Which organ of the alimentary canal has a muscularis externa that contains skeletal muscle?

_____

8. The following questions pertain to bile.

a. What cells produce bile and where are these cells located?

_____

_____

b. Where is bile stored?

_____

_____

c. What is the digestive function of bile?

_____

_____

d. Into which organ of the alimentary canal is bile ejected?

_____

_____

II. Apply What You Learned

Bariatric surgery may be used in cases of extreme obesity to help an individual lose weight. There are different types of bariatric surgeries, gastric bypass being one of the most common.

1. Given the name, what do you think a gastric bypass surgery entails?

   _____

   _____

   _____

   _____

   _____

   _____

   _____

   _____

   _____

2. Would any portion of the stomach still need to be included in the reorganized alimentary canal? Why or why not?

   _____

   _____

   _____

   _____

   _____

   _____

3. To decrease absorption, which portion of the small intestine would you expect to be bypassed or excised in the reorganized alimentary canal?

   _____

   _____

SELF REVIEW | Anatomical Models

Exercise 9.5 Teeth

GO TO ⟩ ANATOMICAL MODELS > DIGESTIVE SYSTEM > SELF REVIEW > IMAGE 1

- *Mouse over the image to locate and label the structures below. Click on the structures to hear their pronunciations.*

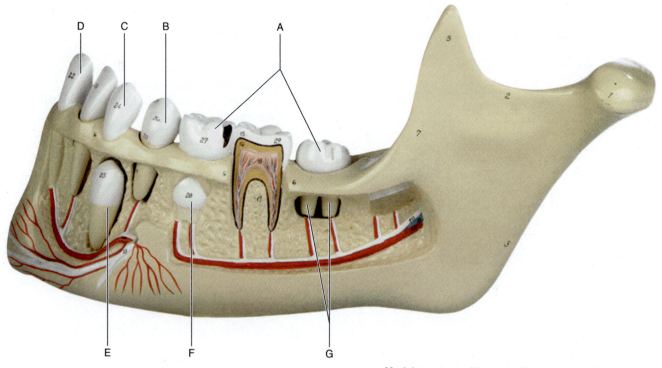

Model courtesy of Denoyer-Geppert, www.denoyer.com

A. _____   E. _____

B. _____   F. _____

C. _____   G. _____

D. _____

Exercise 9.6 **Salivary Glands**

GO TO⟩ ANATOMICAL MODELS > DIGESTIVE SYSTEM > SELF REVIEW > IMAGES 3 AND 4

- *Mouse over the images to locate and label the structures below. Click on the structures to hear their pronunciations.*

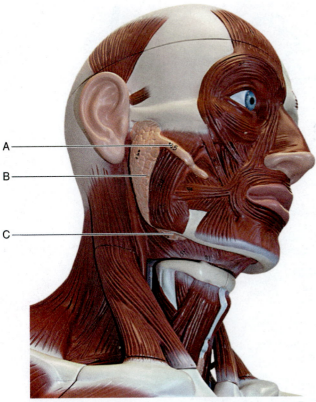

LT-VA16: Life-size muscle torso, 27-part, 3B Scientific®

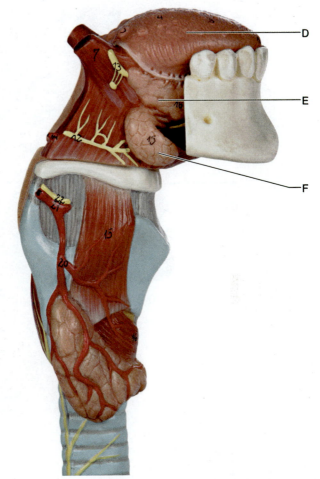

A. _____   D. _____

B. _____   E. _____

C. _____   F. _____

Exercise 9.7 Digestive Structures of the Head and Neck

GO TO ANATOMICAL MODELS > DIGESTIVE SYSTEM > SELF REVIEW > IMAGE 5

• *Mouse over the image to locate and label the structures below. Click on the structures to hear their pronunciations.*

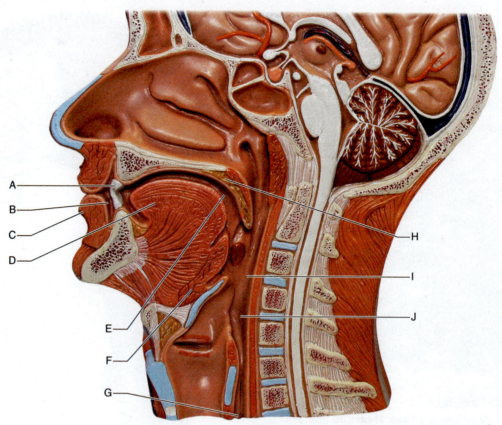

LT-C14: Half head with musculature, 3B Scientific®

A. _____   F. _____

B. _____   G. _____

C. _____   H. _____

D. _____   I. (space) _____

E. (space) _____   J. (space) _____

Exercise 9.8   **Digestive Structures of the Abdominal Cavity**

GO TO ANATOMICAL MODELS > DIGESTIVE SYSTEM > SELF REVIEW > IMAGES 6 AND 7

- *Mouse over the images to locate and label the structures below. Click on the structures to hear their pronunciations.*

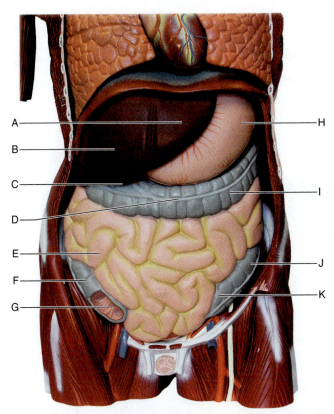

LT-VA16: Life-size muscle torso, 27-part, 3B Scientific®

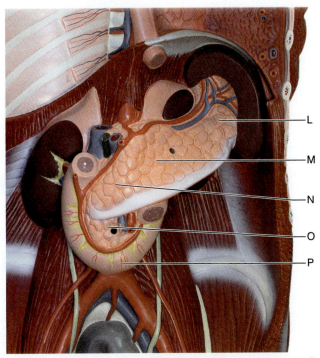

LT-VA16: Life-size muscle torso, 27-part, 3B Scientific®

A. _____   I. _____

B. _____   J. _____

C. _____   K. _____

D. _____   L. _____

E. _____   M. _____

F. _____   N. _____

G. _____   O. _____

H. _____   P. _____

Exercise 9.9   Large Intestine, Posterior View

GO TO ⟩ ANATOMICAL MODELS > DIGESTIVE SYSTEM > SELF REVIEW > IMAGE 9

- *Mouse over the image to locate and label the structures below. Click on the structures to hear their pronunciations.*

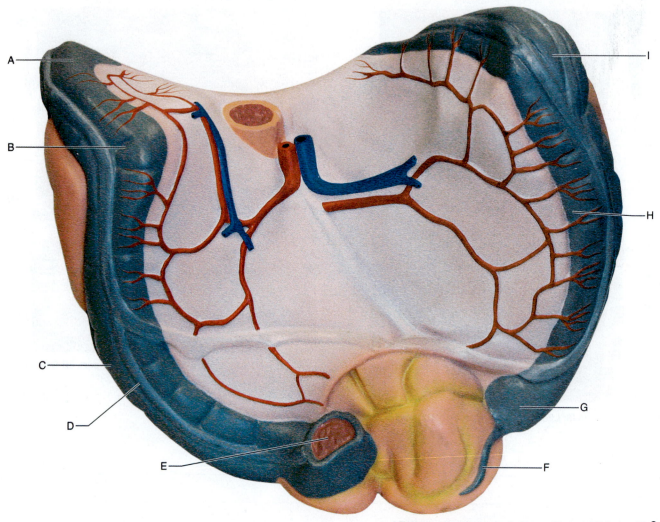

LT-VA16: Life-size muscle torso, 27-part, 3B Scientific®

A. _____   F. _____

B. _____   G. _____

C. _____   H. _____

D. _____   I. _____

E. _____

Exercise 9.10   Stomach and Liver

GO TO ANATOMICAL MODELS > DIGESTIVE SYSTEM > SELF REVIEW > IMAGES 8 AND 11

- *Mouse over the images to locate and label the structures below. Click on the structures to hear their pronunciations.*

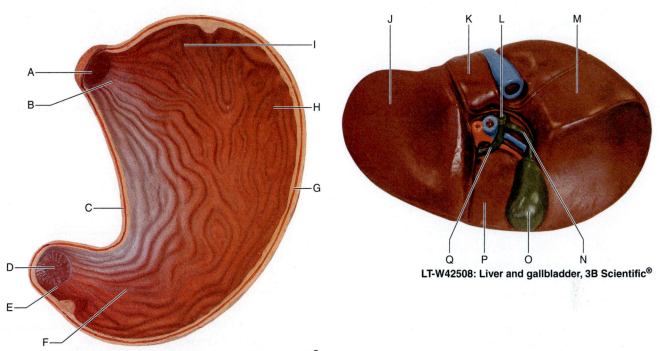

LT-VA16: Life-size muscle torso, 27-part, 3B Scientific®

LT-W42508: Liver and gallbladder, 3B Scientific®

A. _____

B. _____

C. _____

D. _____

E. _____

F. _____

G. _____

H. _____

I. _____

J. _____

K. _____

L. _____

M. _____

N. _____

O. _____

P. _____

Q. _____

QUIZ Anatomical Models

Answer the following questions using information from PAL 3.1 as well as other course materials including your textbook, lecture, and lab notes.

I. Check Your Understanding

1. For each of the different types of teeth listed below, indicate the number of which are found in the lower jaw.

   incisors   _____

   canines   _____

   premolars _____

   molars   _____

2. Identify the large, longitudinal folds of the empty stomach.

   _____

3. What smooth muscle sphincter controls the release of chyme from the stomach?

   _____

4. Which portions of the pharynx are considered part of the digestive system? What type of epithelium lines these regions?

   _____

   _____

   _____

5. For each of the organs listed below, indicate which of the following digestive processes occurs in that organ: **mechanical digestion**, **chemical digestion**, **absorption**, or **none**.

   BEYOND PAL

   oral cavity   _____

   esophagus   _____

   stomach   _____

   small intestine _____

   large intestine _____

   liver   _____

6. Within which organ does the majority of absorption occur?

   _____

7. Which of the accessory digestive organs has the primary role of processing absorbed nutrients?

   _____

8. Lipids are too large for blood capillaries of the alimentary canal to pick up and directly transport. What structures of the lymphatic system absorb lipids?

   _____

   _____

9. The large intestine ends at the anal canal where there are two sphincters. Identify the different sphincters. What type of tissue makes up each of the sphincters? Are they under voluntary or involuntary control?

   _____

   _____

   _____

   _____

   _____

   _____

II. Apply What You Learned

Appendicitis occurs when bacteria become trapped in and infect the wall of the vermiform appendix. This causes an inflammation of the appendix.

1. What organ does the lumen of the appendix open into?

   _____

   _____

2. If inflammation occurs, what happens to the connection between the appendix and the above organ?

   _____

   _____

   _____

   _____

3. If the body cannot fight off the infection, the inflammation could result in a burst appendix. If the appendix bursts, what is the danger to the individual?

   _____

   _____

   _____

   _____

   _____

   _____

4. If the appendix is removed, is there any impact on the overall health of the individual?

   _____

   _____

   _____

   _____

   _____

   _____

TISSUES OF THE DIGESTIVE SYSTEM

SELF REVIEW ⟩ Histology

Exercise 9.11 **Esophagus 40x**

GO TO ⟩ HISTOLOGY > DIGESTIVE SYSTEM > SELF REVIEW > IMAGE 1

- *Mouse over the image to locate and label the structures below. Click on the structures to hear their pronunciations.*

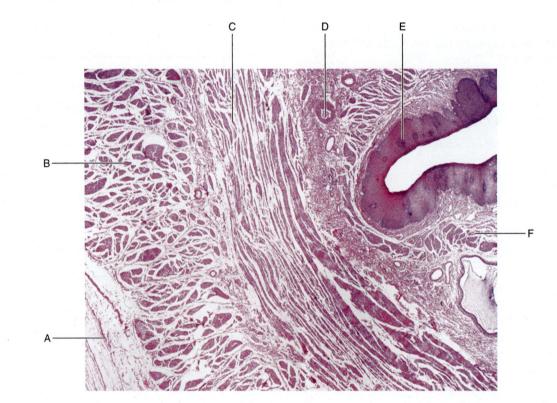

A. _____ D. _____

B. _____ E. _____

C. _____ F. _____

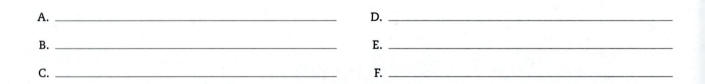

Exercise 9.12 **Stomach 40x and 100x**

GO TO ⟩ HISTOLOGY > DIGESTIVE SYSTEM > SELF REVIEW > IMAGES 7 AND 8

• *Mouse over the images to locate and label the structures below. Click on the structures to hear their pronunciations.*

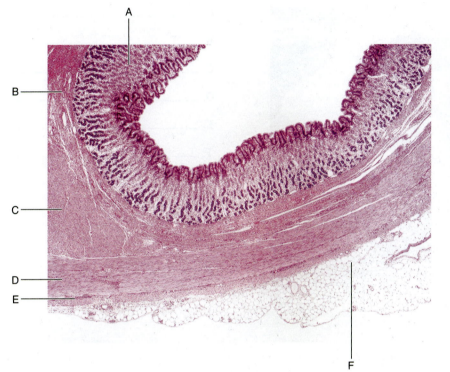

A. _____

B. _____

C. _____

D. _____

E. _____

F. _____

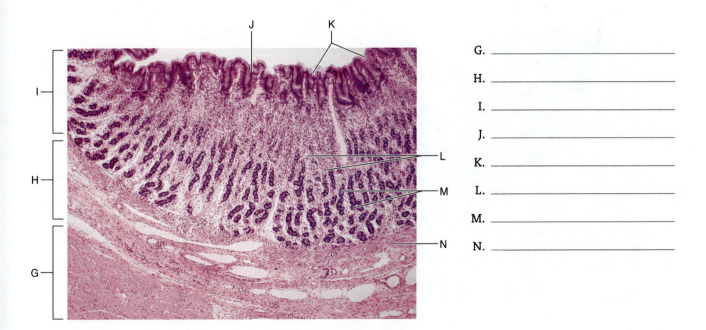

G. _____

H. _____

I. _____

J. _____

K. _____

L. _____

M. _____

N. _____

Exercise 9.13 Duodenum 40x and 100x

GO TO HISTOLOGY > DIGESTIVE SYSTEM > SELF REVIEW > IMAGES 13 AND 15

- *Mouse over the images to locate and label the structures below. Click on the structures to hear their pronunciations.*

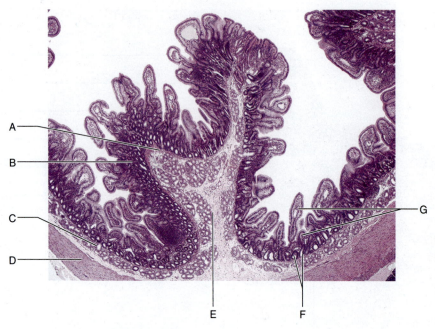

A. _____

B. _____

C. _____

D. _____

E. _____

F. _____

G. _____

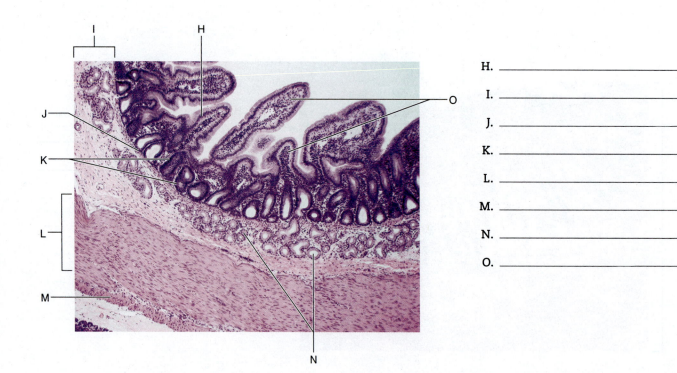

H. _____

I. _____

J. _____

K. _____

L. _____

M. _____

N. _____

O. _____

Jejunum 100x

GO TO HISTOLOGY > DIGESTIVE SYSTEM > SELF REVIEW > IMAGE 22

- *Mouse over the image to locate and label the structures below. Click on the structures to hear their pronunciations.*

A. _____

B. _____

C. _____

D. _____

E. _____

F. _____

G. _____

Ileum 40x

GO TO HISTOLOGY > DIGESTIVE SYSTEM > SELF REVIEW > IMAGE 24

- *Mouse over the image to locate and label the structures below. Click on the structures to hear their pronunciations.*

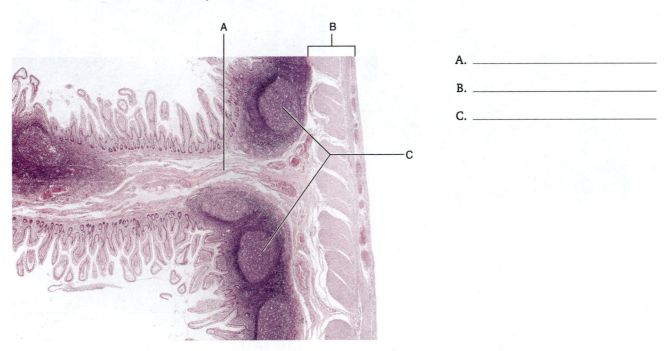

A. _____

B. _____

C. _____

Exercise 9.16 Large Intestine 40x

GO TO HISTOLOGY > DIGESTIVE SYSTEM > SELF REVIEW > IMAGE 27

- *Mouse over the image to locate and label the structures below. Click on the structures to hear their pronunciations.*

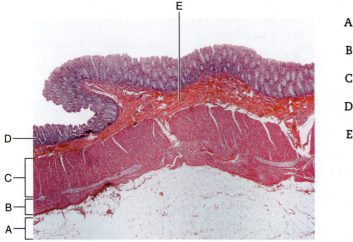

A. _____

B. _____

C. _____

D. _____

E. _____

Exercise 9.17 Submandibular and Sublingual Glands 400x

GO TO HISTOLOGY > DIGESTIVE SYSTEM > SELF REVIEW > IMAGES 34 AND 36

- *Mouse over the images to locate and label the structures below. Click on the structures to hear their pronunciations.*

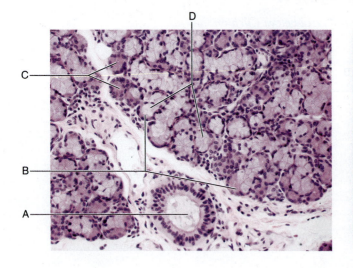

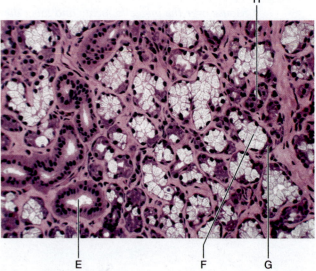

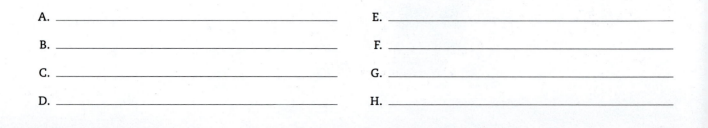

A. _____

B. _____

C. _____

D. _____

E. _____

F. _____

G. _____

H. _____

Exercise 9.18   Parotid Gland 400x

GO TO ⟩ HISTOLOGY > DIGESTIVE SYSTEM > SELF REVIEW > IMAGE 38

- *Mouse over the image to locate and label the structures below. Click on the structures to hear their pronunciations.*

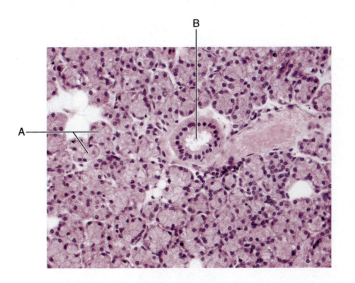

A. _____

B. _____

Exercise 9.19   Pancreas 100x

GO TO ⟩ HISTOLOGY > DIGESTIVE SYSTEM > SELF REVIEW > IMAGE 40

- *Mouse over the image to locate and label the structures below. Click on the structures to hear their pronunciations.*

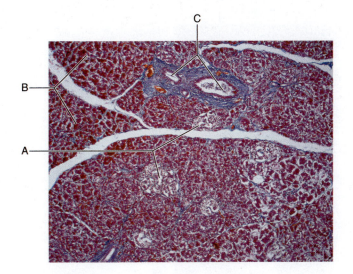
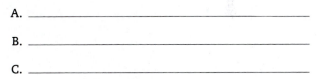

A. _____

B. _____

C. _____

Exercise 9.20   Liver 100x and 200x

GO TO ▷ HISTOLOGY > DIGESTIVE SYSTEM > SELF REVIEW > IMAGES 47 AND 48

- *Mouse over the images to locate and label the structures below. Click on the structures to hear their pronunciations.*

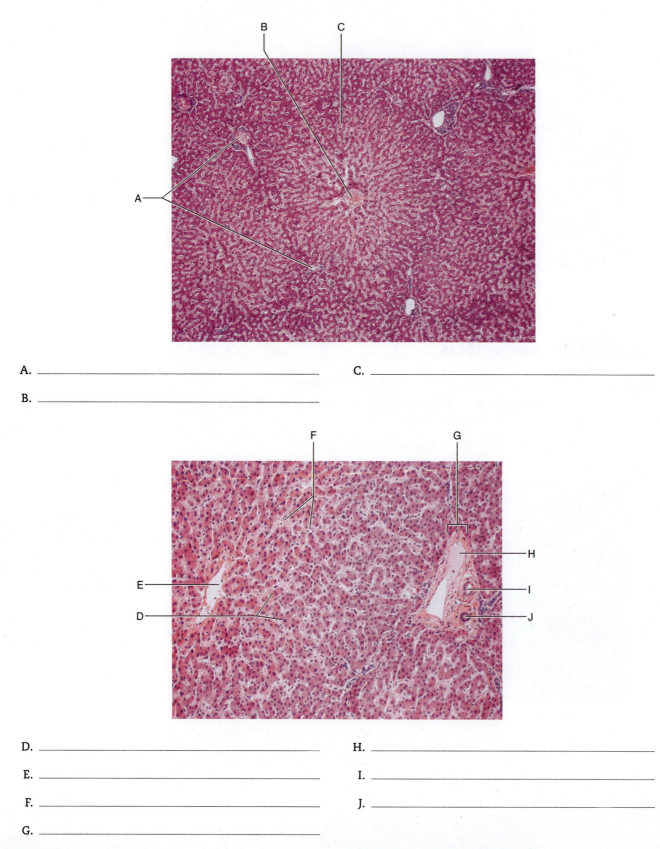

A. _____   C. _____

B. _____

D. _____   H. _____

E. _____   I. _____

F. _____   J. _____

G. _____

QUIZ │ Histology

Answer the following questions using information from PAL 3.1 as well as other course materials including your textbook, lecture, and lab notes.

I. Check Your Understanding

1. Which of the salivary glands is composed of serous acini, but has no mucous acini?

   _____

2. The small intestine has three major adaptations that combine to drastically increase the overall surface area. Name them.

   _____

   _____

3. The muscularis externa is typically composed of two smooth muscle layers. Name them.

   _____

   _____

4. The muscularis externa layer of the stomach has a third smooth muscle layer. Identify this third layer.

   _____

5. Indicate the type of epithelium lining the lumen of each of the following organs.

   oral cavity _____

   esophagus _____

   stomach _____

   small intestine _____

   large intestine _____

6. What structures make up a portal triad of a liver lobule?

   _____

   _____

BEYOND PAL 7. What is the function of goblet cells?

   _____

   _____

8. Which structure of a portal triad transports oxygen-rich blood?

   _____

9. Which structure of a portal triad transports nutrient-rich, oxygen-poor blood?

   _____

10. Why are BOTH answers for questions 8 and 9 necessary for hepatocytes to function?

   _____

   _____

   _____

   _____

   _____

   _____

   _____

11. Match the following cells or structures, at left, to their correct function, at right.

   _____ chief cells

   _____ parietal cells

   _____ hepatocytes

   _____ Peyer's patches

   _____ duodenal glands

   _____ serous acinus

   _____ intestinal crypts

   a. produce alkaline mucus to neutralize chyme

   b. secrete intestinal juice

   c. secrete pepsinogen which, when converted to pepsin, breaks down proteins

   d. lymphoid tissue that combats pathogens trying to enter through intestine wall

   e. secretes enzymes into saliva

   f. store sugar as glycogen

   g. secrete hydrochloric acid which converts pepsinogen to pepsin

II. Apply What You Learned

Peptic ulcers can occur in regions of the alimentary canal that are highly acidic. They can be very painful, as acid breaks down the wall of the affected organ, and they can lead to bleeding and in extreme cases even death.

1. Given that peptic ulcers tend to occur in organs that are transporting acidic contents, which organs would you predict to be most prone to peptic ulcers?

_____

_____

2. Peptic ulcers can occur in the esophagus. How might this happen, and is the esophagus set up to handle acidic contents?

_____

_____

_____

_____

_____

_____

_____

_____

_____

3. One common cause of peptic ulcers is the use of certain medications that inhibit the secretion of mucus. Why might this contribute to the development of ulcers?

_____

_____

_____

_____

_____

_____

_____

4. Another common cause of peptic ulcers is inflammation caused by *Helicobacter pylori* bacteria, which can colonize the stomach or intestine. Typically, these bacteria impact the production of gastrin. What is the function of gastrin, where is it produced, and how would too much gastrin be bad?

_____

_____

_____

_____

_____

_____

_____

_____

LAB PRACTICAL Digestive System

1. Identify the structure.

   _____

2. Identify the structure.

   _____

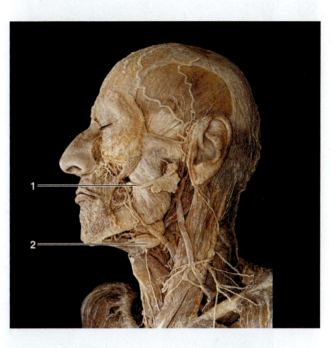

3. Identify the type of tooth.

   _____

4. How many **total** teeth of this type are found in permanent adult dentition?

   _____

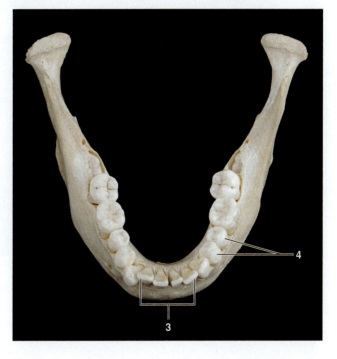

LAB PRACTICAL *continues*

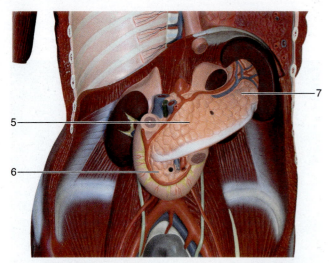

LT-VA16: Life-size muscle torso, 27-part, 3B Scientific®

5. Identify this part of the pancreas.

_____

6. Identify this part of the pancreas.

_____

7. Identify this part of the pancreas.

_____

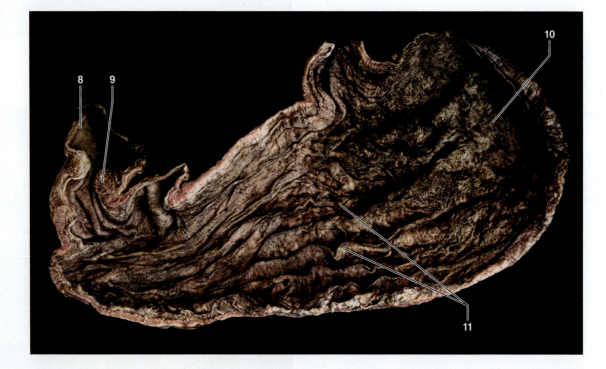

8. Identify the structure.

_____

9. Identify the structure.

_____

10. Identify the region.

_____

11. Identify the structures.

_____

12. Identify the structure.

_____

13. Identify the structure.

_____

14. Identify the structure.

_____

15. Identify the structure.

_____

16. Identify the structure.

_____

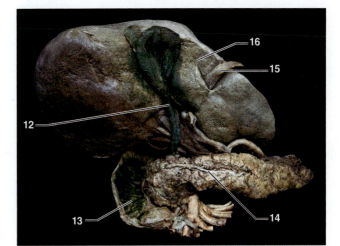

17. Identify the structure.

_____

18. Identify the structure.

_____

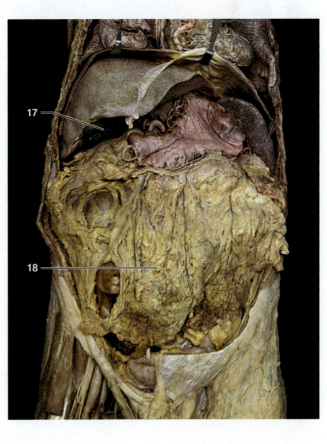

LAB PRACTICAL _continues_

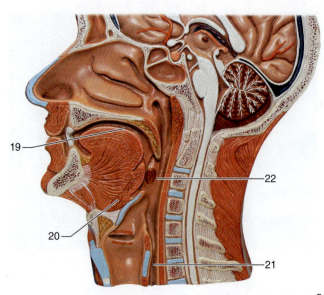

LT-C14: Half head with musculature, 3B Scientific®

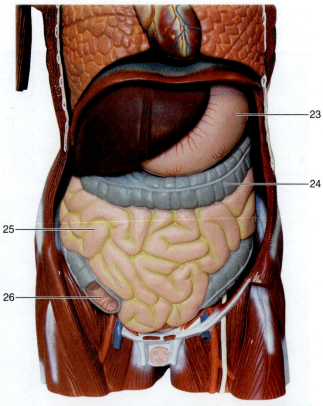

LT-VA16: Life-size muscle torso, 27-part, 3B Scientific®

19. Identify the space.

_____

20. Identify the structure.

_____

21. Identify the structure.

_____

22. Identify the space.

_____

23. Identify the structure.

_____

24. Identify the region of the large intestine.

_____

25. Identify the structure.

_____

26. Identify the structure.

_____

27. Identify the region of the large intestine.

_____

28. Identify the structure of the large intestine.

_____

29. Identify the structure.

_____

30. Identify the region of the large intestine.

_____

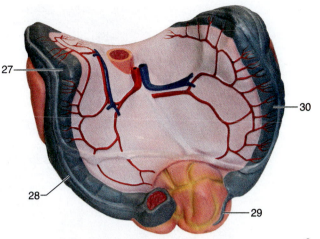

LT-VA16: Life-size muscle torso, 27-part, 3B Scientific®

31. Identify the type of epithelium shown in the image at right.

_____

32. Identify the layer.

_____

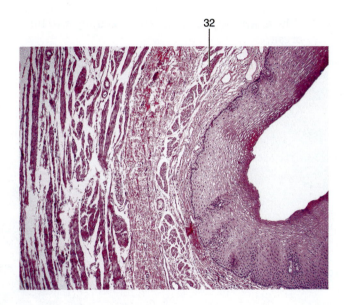

LAB PRACTICAL *continues*

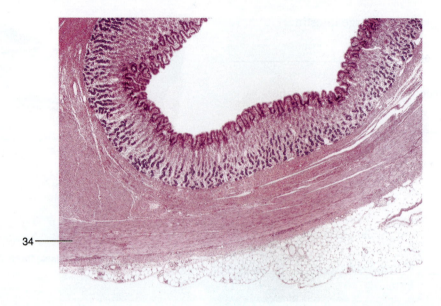

33. Which layer in the image above is unique to this organ?

_____

34. Identify the type of tissue.

_____

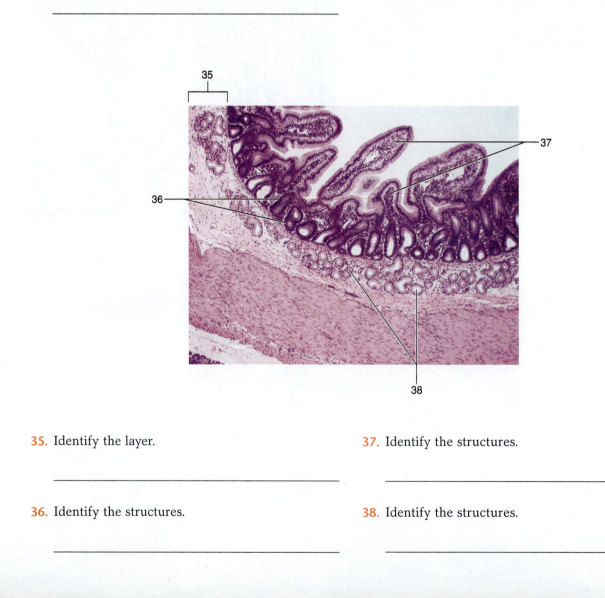

35. Identify the layer.

_____

36. Identify the structures.

_____

37. Identify the structures.

_____

38. Identify the structures.

_____

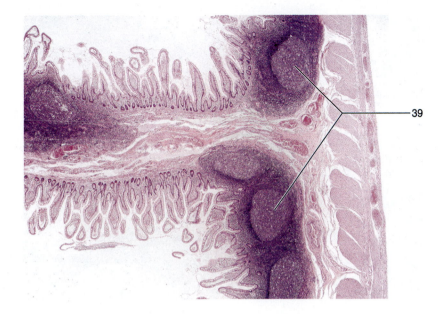

39. Identify the structures.

_____

40. Identify the region of the small intestine in the image above.

_____

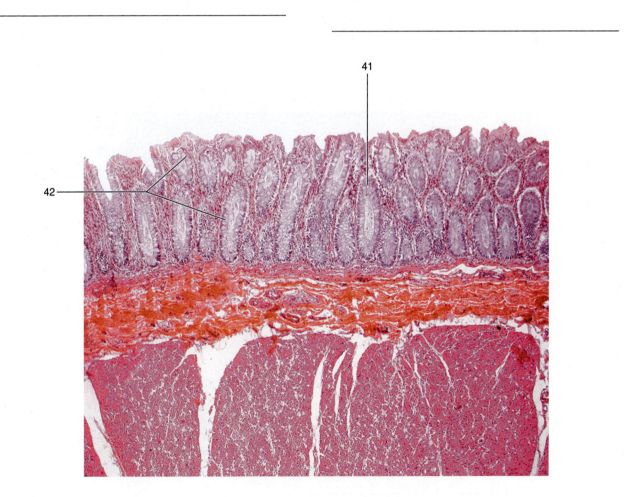

41. Identify the epithelium.

_____

42. Identify the cells.

_____

LAB PRACTICAL _continues_

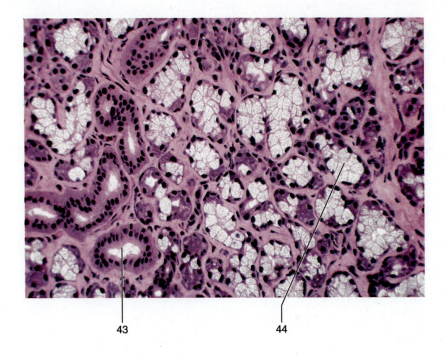

43   44

43. Identify the structure.

_____

44. Identify the structure.

_____

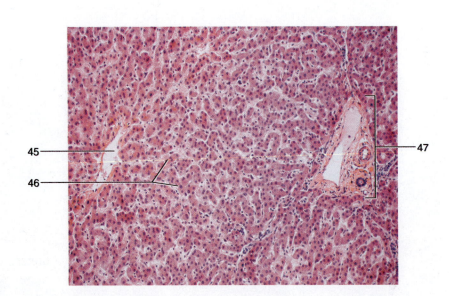

45 ———

46 ———

——— 47

45. Identify the structure.

_____

47. Identify the structure.

_____

46. Identify the cell type.

_____

CHAPTER 10

The Urinary System

STUDENT OBJECTIVES

GROSS ANATOMY OF THE URINARY SYSTEM

1. Understand the location, coverings, and external gross anatomy of the kidney.
2. Understand blood flow to and within the kidney.
3. Understand the organization of the renal cortex and the renal medulla.
4. Identify the major structures of the kidney.
5. Trace the formation of urine from renal corpuscle to renal papilla.
6. Trace the flow of formed urine from the renal papilla to where it leaves the urethra.
7. Understand the location and function of the ureters.

8. Understand the location and function of the urinary bladder.
9. Identify the female urethra and the different regions of the male urethra.

TISSUES OF THE URINARY SYSTEM

10. Identify the regions of a nephron and understand the role each plays in the formation of urine.
11. Identify the layers of the ureters, urinary bladder, and urethra.
12. Identify the epithelial types that line each region of the urinary system.
13. Understand the histological organization of each organ of the urinary system.

GROSS ANATOMY OF THE URINARY SYSTEM

SELF REVIEW | Human Cadaver

Exercise 10.1   Urinary System

GO TO > HUMAN CADAVER > URINARY SYSTEM > SELF REVIEW > IMAGE 5

- *Mouse over the image to locate and label the structures indicated below. Click on the structures to hear their pronunciations.*

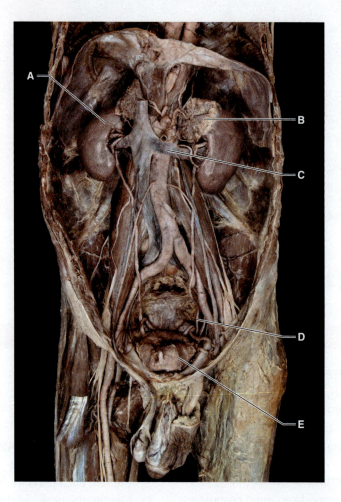

A. _____   D. _____

B. _____   E. _____

C. _____

Exercise 10.2 **Kidney**

GO TO HUMAN CADAVER > URINARY SYSTEM > SELF REVIEW > IMAGE 4

- *Mouse over the image to locate and label the structures indicated below. Click on the structures to hear their pronunciations.*

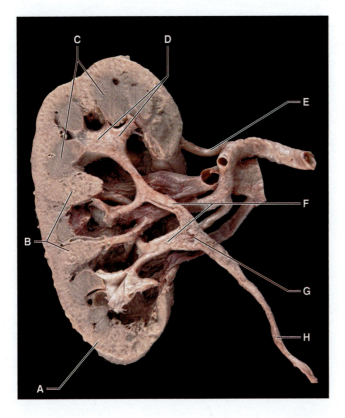

A. _____

B. _____

C. _____

D. _____

E. _____

F. _____

G. _____

H. _____

QUIZ Human Cadaver

Answer the following questions using information from PAL 3.1 as well as other course materials including your textbook, lecture, and lab notes.

I. Check Your Understanding

1. Arrange the structures at right in the order that urine flows, starting from where it exits a medullary pyramid (1) and ending where it leaves the body (7).

1. _____   a. ureter

2. _____   b. minor calyx

3. _____   c. major calyx

4. _____   d. renal papilla

5. _____   e. urethra

6. _____   f. renal pelvis

7. _____   g. urinary bladder

2. What are the portions of the renal cortex that extend between the renal pyramids?

_____

_____

3. Which kidney is typically slightly lower in the abdominal cavity compared to the other?

_____

BEYOND PAL

4. What is the tough fibrous layer surrounding each kidney?

_____

5. Kidneys are retroperitoneal. What does that mean?

_____

_____

_____

_____

II. Apply What You Learned

Kidney stones are solid masses made up of tiny crystals, and can be found in the kidney or the ureters. The crystals can start to form when the urine contains a high enough concentration of a substance that it starts to precipitate. Calcium stones are the most common.

1. What space in the kidney could contain a kidney stone? Why?

_____

_____

_____

_____

2. Why would drinking more fluids decrease the chance of a stone forming?

_____

_____

_____

_____

3. Kidney stones can be quite painful. What causes the pain?

_____

_____

_____

_____

_____

_____

4. As a kidney stone moves through the ureter, the pain associated with it can come and go. Why would the pain stop and then start again?

_____

_____

_____

_____

_____

_____

SELF REVIEW | Anatomical Models

Exercise 10.3   Kidney

GO TO › ANATOMICAL MODELS > URINARY SYSTEM > SELF REVIEW > IMAGE 1

- *Mouse over the image to locate and label the structures indicated below. Click on the structures to hear their pronunciations.*

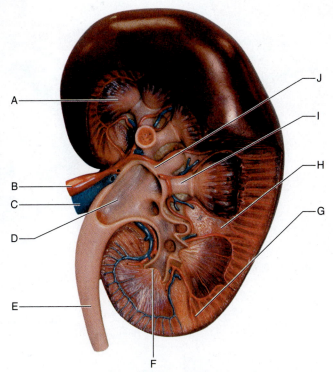

Copyright by SOMSO, 2010, www.somso.com

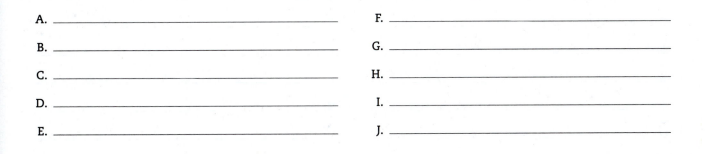

A. _____   F. _____

B. _____   G. _____

C. _____   H. _____

D. _____   I. _____

E. _____   J. _____

Exercise 10.4 Nephron

ANATOMICAL MODELS > URINARY SYSTEM > SELF REVIEW > IMAGES 7 AND 8

- *Mouse over the images to locate and label the structures indicated below. Click on the structures to hear their pronunciations.*

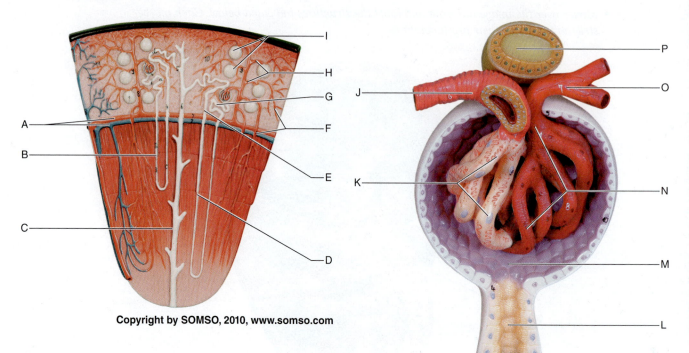

Copyright by SOMSO, 2010, www.somso.com

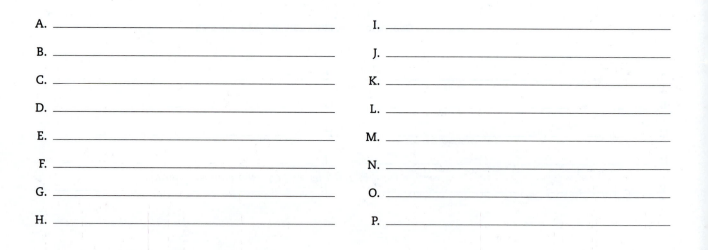

Copyright by SOMSO, 2010, www.somso.com

A. _____

B. _____

C. _____

D. _____

E. _____

F. _____

G. _____

H. _____

I. _____

J. _____

K. _____

L. _____

M. _____

N. _____

O. _____

P. _____

Exercise 10.5 **Male and Female Urinary Systems**

GO TO ANATOMICAL MODELS > URINARY SYSTEM > SELF REVIEW > IMAGES 10 AND 14

- *Mouse over the images to locate and label the structures indicated below. Click on the structures to hear their pronunciations.*

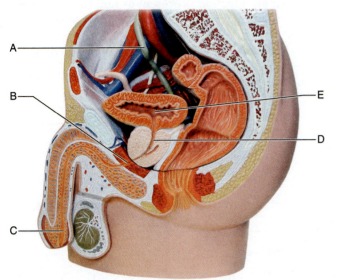

LT-VG351: Male pelvis, 2-part, 3B Scientific®

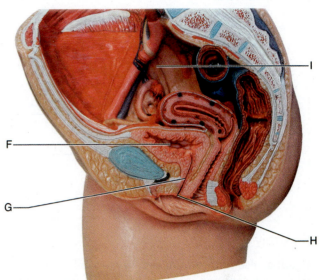

LT-H10: Female pelvis, 2-part, 3B Scientific®

A. _____

B. _____

C. _____

D. _____

E. _____

F. _____

G. _____

H. _____

I. _____

QUIZ | Anatomical Models

Answer the following questions using information from PAL 3.1 as well as other course materials including your textbook, lecture, and lab notes.

I. Check Your Understanding

1. Which segment of a nephron is contained within the renal medulla?

_____

2. Which arteries are found at the border between a renal pyramid and the cortex?

_____

_____

3. Which layer of the kidney contains the renal corpuscles?

_____

4. The portion of the male urethra that passes through the penis is contained in which penile tissue?

_____

5. For each of the following regions of a uriniferous tubule, indicate which process

5. For each of the following regions of a uriniferous tubule, indicate which process it is involved in. Choose from **secretion**, **resorption**, and **filtration**. (List all that apply.)

glomerular
capillaries

_____

descending loop
of Henle

_____

ascending loop
of Henle (thick
segment)

_____

proximal
convoluted tubule

_____

distal convoluted
tubule

_____

collecting duct

_____

II. Apply What You Learned

Cystitis, also known as a urinary tract infection (UTI), is a common infection seen most frequently in females. These infections typically originate in the urinary bladder, but can spread to the ureters and kidneys.

1. What anatomical difference between males and females is responsible for the higher incidence of cystitis in females? How does this difference contribute to the infection?

_____

_____

_____

_____

_____

_____

2. In patients displaying symptoms of a UTI, a urine sample is typically requested so the urine can be examined under a microscope, or to conduct other tests. The number of white blood cells, red blood cells, and bacteria in a defined area is very useful in determining if an infection is present or not. Compared to normal state, would you expect to see higher, lower, or similar counts of each of the following cell types when an infection is present? Why?

white blood cells

_____

_____

red blood cells

_____

_____

bacteria

_____

_____

3. Why is it painful to urinate (empty the bladder) when a UTI is present?

_____

_____

_____

_____

TISSUES OF THE URINARY SYSTEM

SELF REVIEW | Histology

Exercise 10.6 **Kidney, Cortex and Medulla 400x**

GO TO HISTOLOGY > URINARY SYSTEM > SELF REVIEW > IMAGES 3 AND 5

- *Mouse over the images to locate and label the structures indicated below. Click on the structure to hear the pronunciation.*

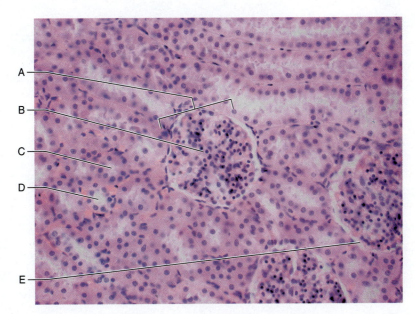

A. _____   D. _____

B. _____   E. _____

C. _____

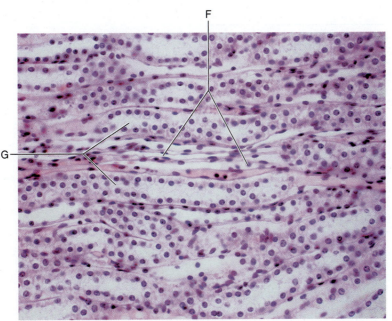

F. _____   G. _____

Urinary Bladder 100x

GO TO   HISTOLOGY > URINARY SYSTEM > SELF REVIEW > IMAGE 10

- *Mouse over the image to locate and label the structures indicated below. Click on the structures to hear their pronunciations.*

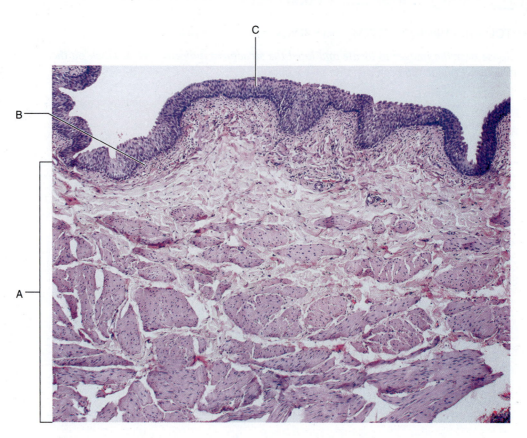

A. _____

B. _____

C. _____

Exercise 10.8 Ureter 100x and 400x

GO TO HISTOLOGY > URINARY SYSTEM > SELF REVIEW > IMAGES 15 AND 16

- *Mouse over the images to locate and label the structures indicated below. Click on the structures to hear their pronunciations.*

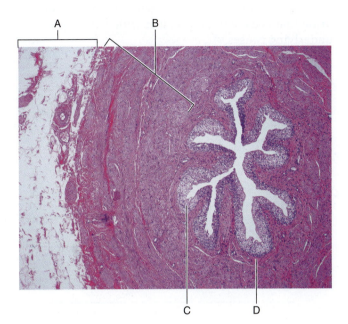
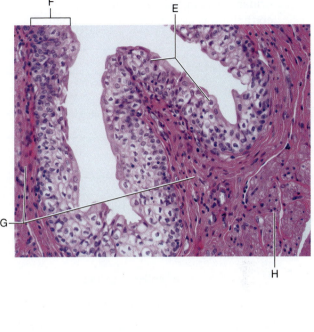

A. _____

B. _____

C. _____

D. _____

E. _____

F. _____

G. _____

H. _____

QUIZ | Histology

Answer the following questions using information from PAL 3.1 as well as other course materials including your textbook, lecture, and lab notes.

I. Check Your Understanding

1. For each of the following structures of the urinary system, indicate which type of epithelium is associated with its lumen.

   ureter _____

   urinary bladder _____

   proximal
   convoluted tubule _____

   collecting duct _____

   descending loop
   of Henle _____

2. Arrange the structures at right in order of filtrate transport, starting where filtrate enters a nephron (1), and ending where urine leaves the uriniferous tubule (7).

   1. _____   a. collecting duct

   2. _____   b. descending loop of Henle

   3. _____   c. proximal convoluted tubule

   4. _____   d. glomerulus

   5. _____   e. distal convoluted tubule

   6. _____   f. ascending loop of Henle

   7. _____   g. minor calyx

3. Which type of muscle tissue is found in the following locations/structures:

   ureter _____

   urinary bladder
   (detrusor muscle) _____

   internal urethral
   sphincter _____

   external urethral
   sphincter _____

4. Which region(s) of the male urethra contain(s) transitional epithelium?

   _____

   _____

II. Apply What You Learned

Urinary incontinence is an involuntary leakage of urine that is a common problem with a variety of underlying causes.

1. Which muscles control the release of urine from the bladder?

   _____

   _____

2. Which muscle is involuntary? What activates this muscle?

   _____

   _____

   _____

   _____

3. Which muscle is voluntary? What type of muscle tissue is this?

   _____

   _____

4. Stress incontinence is the most common type of incontinence. It is characterized as leaking urine during a physical activity; while coughing, sneezing, or laughing; or during the later stages of pregnancy. What do you think causes of this type of incontinence?

   _____

   _____

   _____

   _____

   _____

   _____

   _____

LAB PRACTICAL Urinary System

1. Identify the structures.

    _____

2. Identify the structure.

    _____

3. Identify the structures.

    _____

4. Identify the structure.

    _____

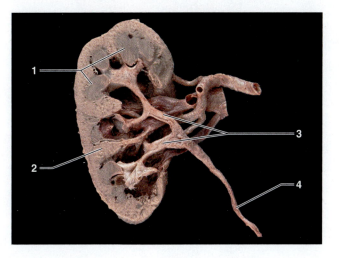

5. Identify the structures.

    _____

6. Identify the structure.

    _____

7. Identify the structure.

    _____

8. Identify the structure.

    _____

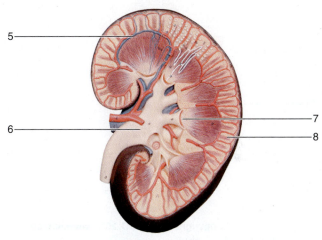

Copyright by SOMSO, 2010, www.somso.com

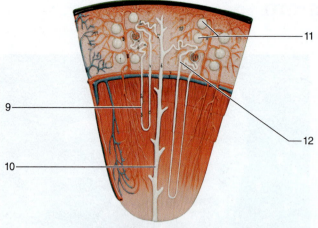

9. Identify the structure (be specific).

_____

10. Identify the structure.

_____

11. Identify the structures.

_____

12. Identify the structure.

_____

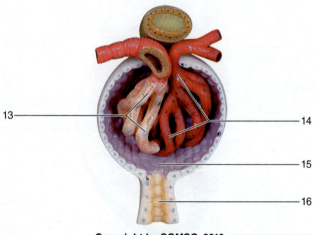

13. Identify the cells.

_____

14. Identify the structures.

_____

15. Identify the structure.

_____

16. Identify the structure.

_____

17. Identify the structure.

_____

18. Identify the structure (be specific with region).

_____

19. Identify the organ.

_____

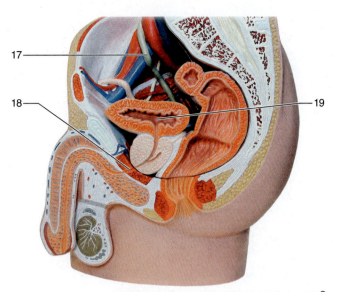

LT-VG351: Male pelvis, 2-part, 3B Scientific®

20. Identify the structure.

_____

21. Identify the structure.

_____

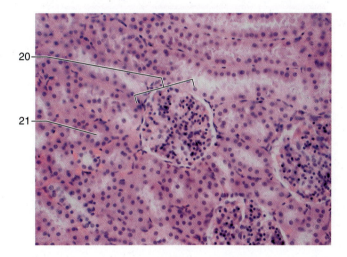

22. Identify the tissue.

_____

23. Identify the layer.

_____

24. Identify the layer.

_____

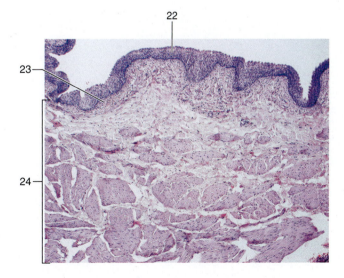

LAB PRACTICAL *continues*

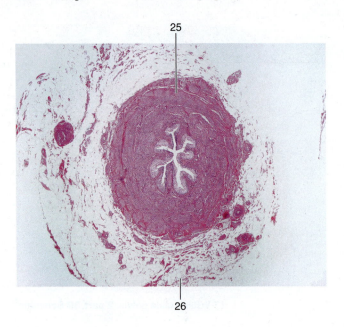

25

26

25. Identify the layer.

_____

26. Identify the layer.

_____

27. Identify the organ shown in the image at left.

_____

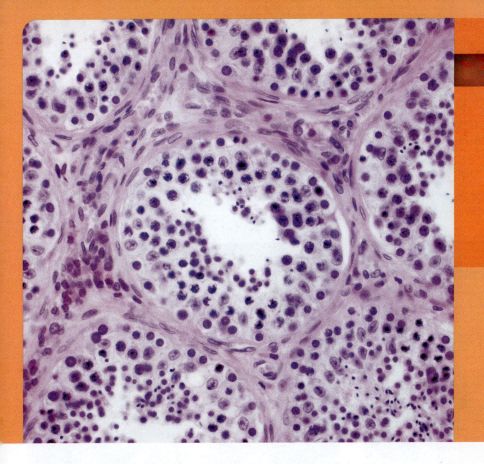

The Reproductive System

STUDENT OBJECTIVES

GROSS ANATOMY OF THE REPRODUCTIVE SYSTEM

1. Identify and name the structures of the male and female reproductive systems.
2. Describe the structure of the testes.
3. Describe the location, structure, and function of the accessory reproductive organs of the male.
4. Trace the pathway of a sperm from its site of origin to the external environment.
5. Describe the location, structure, and function of the ovaries.
6. Describe the location, structure, and function of the organs of the female reproductive duct system.

7. Describe the anatomy of the female external genitalia.
8. Identify the names and origins of the secretory products of the testes and ovaries.

TISSUES OF THE REPRODUCTIVE SYSTEM

9. Describe the histology of the testis, sperm, ovary, uterus, and vagina.
10. Describe the events that take place during spermatogenesis.
11. Describe the process of follicular maturation.
12. Define endometrium and myometrium.

GROSS ANATOMY OF THE REPRODUCTIVE SYSTEM

SELF REVIEW | Human Cadaver

Exercise 11.1 **Penis and Testis, Lateral View**

GO TO ▷ HUMAN CADAVER > REPRODUCTIVE SYSTEM > SELF REVIEW > IMAGES 1, 2, AND 4

- *Mouse over the images to locate and label the structures indicated below. Click on the structures to hear their pronunciations.*

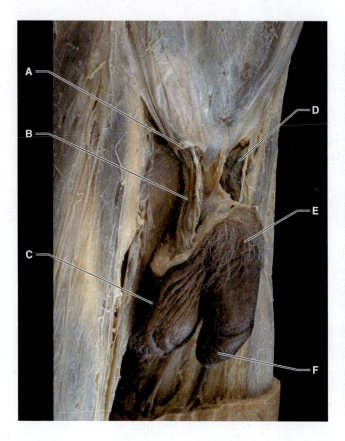
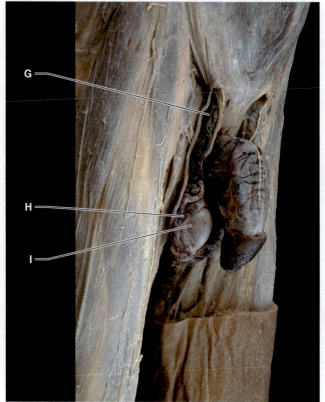

A. _____

B. _____

C. _____

D. _____

E. _____

F. _____

G. _____

H. _____

I. _____

Exercise 11.2 **Male Reproductive Structures and Structures of the Penis**

GO TO › HUMAN CADAVER > REPRODUCTIVE SYSTEM > SELF REVIEW > IMAGES 6 AND 7

> • *Mouse over the images to locate and label the structures indicated below. Click on the structures to hear their pronunciations.*

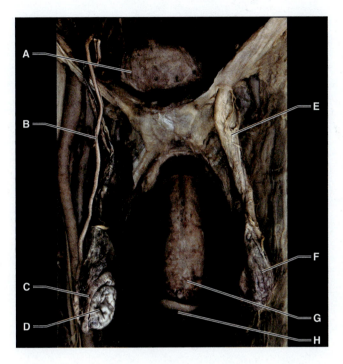

A. _____

B. _____

C. _____

D. _____

E. _____

F. _____

G. _____

H. _____

I. _____

J. _____

Exercise 11.3 **Male Reproductive Structures, Sagittal View**

GO TO HUMAN CADAVER > REPRODUCTIVE SYSTEM > SELF REVIEW > IMAGE 8

- *Mouse over the image to locate and label the structures indicated below. Click on the structures to hear their pronunciations.*

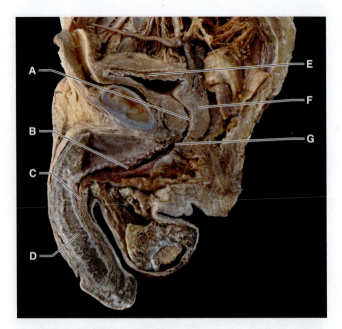

A. _____

B. _____

C. _____

D. _____

E. _____

F. _____

G. _____

Exercise 11.4 **Female External Genitalia**

GO TO HUMAN CADAVER > REPRODUCTIVE SYSTEM > SELF REVIEW > IMAGE 9

- *Mouse over the image to locate and label the structures indicated below. Click on the structures to hear their pronunciations.*

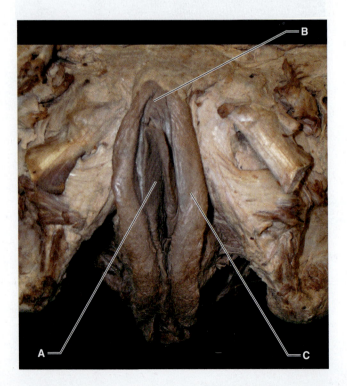

A. _____

B. _____

C. _____

Exercise 11.5 **Uterus**

GO TO ⟩ HUMAN CADAVER > REPRODUCTIVE SYSTEM > SELF REVIEW > IMAGE 10

- *Mouse over the image to locate and label the structures indicated below. Click on the structures to hear their pronunciations.*

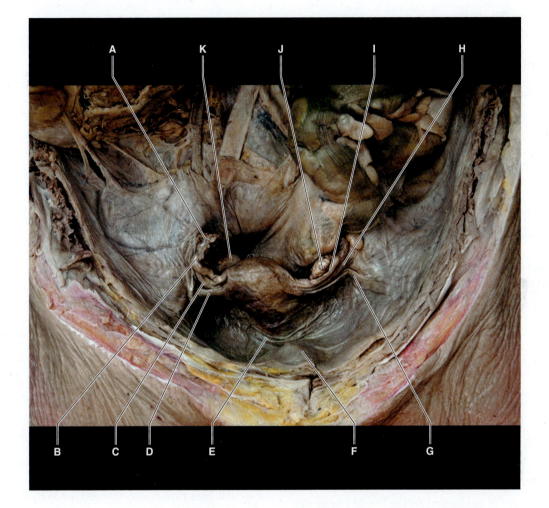

A. _____   G. _____

B. _____   H. _____

C. _____   I. _____

D. _____   J. _____

E. _____   K. _____

F. _____

QUIZ | Human Cadaver

Answer the following questions using information from PAL 3.1 as well as other course materials including your textbook, lecture, and lab notes.

I. Check Your Understanding

1. Arrange the structures at right according to the pathway of sperm. Start with its origin in the testis (1) and end with the external environment (7).

   1. _____   a. membranous urethra

   2. _____   b. ductus deferens

   3. _____   c. prostatic urethra

   4. _____   d. epididymis

   5. _____   e. spongy urethra

   6. _____   f. seminiferous tubule

   7. _____   g. ejaculatory duct

2. What is the name of the tough fibrous layer that surrounds both the testes and the ovaries?

   _____

3. What is the name of the complex venous network that arises from the testicular veins?

   _____

4. What is the name of the fingerlike projections of the uterine tubes that caress each ovary?

   _____

5. The ovary is connected to the uterine wall via which ligament?

   _____

6. Which structure of the female is homologous to the penis in the male?

   _____

   _____

BEYOND PAL 7. What are the two major functions of the ovaries?

   _____

   _____

   _____

   _____

II. Apply What You Learned

A vasectomy is a surgical form of male birth control and a very common procedure. The purpose is to prevent the release of sperm from the male reproductive tract.

1. In a vasectomy, which structure of the male reproductive tract is cut?

   _____

2. Does this procedure prevent the production of sperm? Why or why not?

   _____

   _____

   _____

   _____

3. How are sperm affected by this procedure? What happens to them?

   _____

   _____

   _____

   _____

4. Is this procedure reversible? Why or why not?

   _____

   _____

   _____

   _____

   _____

   _____

SELF REVIEW | Anatomical Models

Exercise 11.6 **Reproductive Organs of the Male**

GO TO > ANATOMICAL MODELS > REPRODUCTIVE SYSTEM > SELF REVIEW > IMAGE 1

- *Mouse over the image to locate and label the structures indicated below. Click on the structures to hear their pronunciations.*

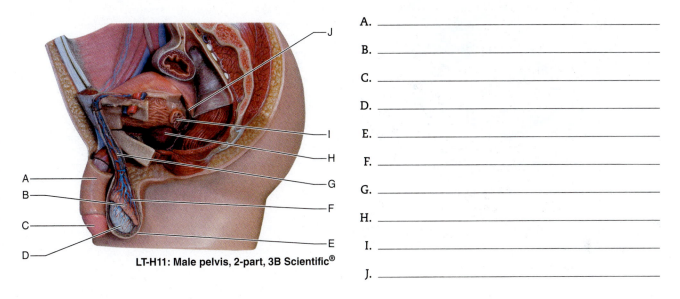

LT-H11: Male pelvis, 2-part, 3B Scientific®

A. _____

B. _____

C. _____

D. _____

E. _____

F. _____

G. _____

H. _____

I. _____

J. _____

Exercise 11.7 **Reproductive Organs of the Male**

GO TO > ANATOMICAL MODELS > REPRODUCTIVE SYSTEM > SELF REVIEW > IMAGE 4

- *Mouse over the image to locate and label the structures indicated below. Click on the structures to hear their pronunciations.*

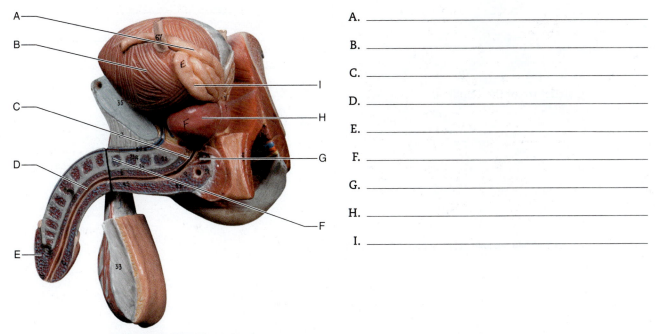

Copyright by SOMSO, 2010, www.somso.com

A. _____

B. _____

C. _____

D. _____

E. _____

F. _____

G. _____

H. _____

I. _____

Exercise 11.8 **Reproductive Organs of the Female**

GO TO ⟩ ANATOMICAL MODELS > REPRODUCTIVE SYSTEM > SELF REVIEW > IMAGE 5

- *Mouse over the image to locate and label the structures indicated below. Click on the structures to hear their pronunciations.*

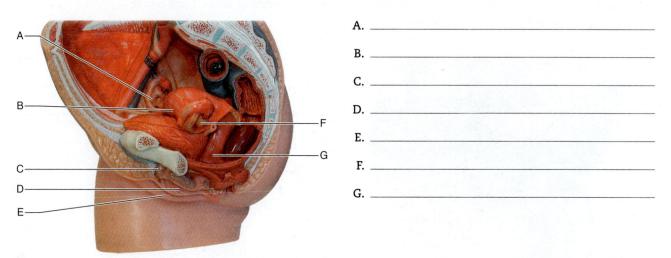

LT-H10: Female pelvis, 2-part, 3B Scientific®

A. _____

B. _____

C. _____

D. _____

E. _____

F. _____

G. _____

Exercise 11.9 **Reproductive Organs of the Female, Deep View**

GO TO ⟩ ANATOMICAL MODELS > REPRODUCTIVE SYSTEM > SELF REVIEW > IMAGE 7

- *Mouse over the image to locate and label the structures indicated below. Click on the structures to hear their pronunciations.*

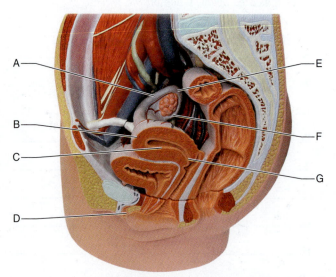

LT-VG366: Female pelvis, 2-part, 3B Scientific®

A. _____

B. _____

C. _____

D. _____

E. _____

F. _____

G. _____

Exercise 11.10 **Mammary Gland**

GO TO ANATOMICAL MODELS > REPRODUCTIVE SYSTEM > SELF REVIEW > IMAGE 12

• *Mouse over the image to locate and label the structures indicated below. Click on the structures to hear their pronunciations.*

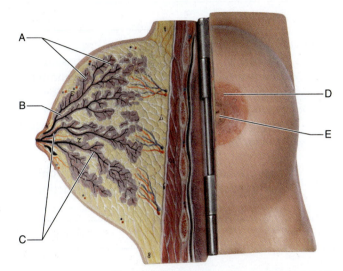

A. _____

B. _____

C. _____

D. _____

E. _____

Copyright by SOMSO, 2010, www.somso.com

QUIZ Anatomical Models

Answer the following questions using information from PAL 3.1 as well as other course materials including your textbook, lecture, and lab notes.

I. Check Your Understanding

1. What important structure runs through the corpus spongiosum?

_____

2. The pampiniform plexus of the testicular vein surrounds which other vessel?

_____

3. Which organ of the male reproductive system is located inferior to the urinary bladder? Which important urinary structure passes through this organ?

_____

4. Which part of the uterus projects into the vagina?

_____

5. Identify the internal reproductive organs (internal genitalia) of the female, excluding the primary sex organs.

_____

_____

6. Identify the ring of pigmented skin that surrounds the nipple.

_____

7. Identify the superior and rounded part of the uterus into which the uterine tubes empty.

_____

_____

_____

_____

BEYOND
PAL

8. Identify the erectile tissues of the penis. Describe the function of these tissues.

_____

_____

_____

_____

II. Apply What You Learned

1. Uterine prolapse is a condition in which the cervix of the uterus descends into the vagina.

a. What is the main cause of uterine prolapse?

_____

_____

_____

_____

_____

_____

_____

b. Identify the structures that support the uterus.

_____

_____

_____

_____

_____

c. What is the basic function of the pelvic floor?

_____

_____

_____

_____

2. The prostate gland is located at the base of the urinary bladder and consists of glandular tissue mixed with smooth muscle and fibrous tissue. Hyperplasia of the prostate (benign prostatic hyperplasia) affects a high proportion of older males. This condition results in constriction of which structure running through the prostate gland? What function is compromised as a result of this constriction?

_____

_____

_____

_____

TISSUES OF THE REPRODUCTIVE SYSTEM

SELF REVIEW | Histology

Exercise 11.11 **Sperm Smear 1000x**

GO TO HISTOLOGY > REPRODUCTIVE SYSTEM > SELF REVIEW > IMAGE 2

• *Mouse over the image to locate and label the structures indicated below. Click on the structures to hear their pronunciations.*

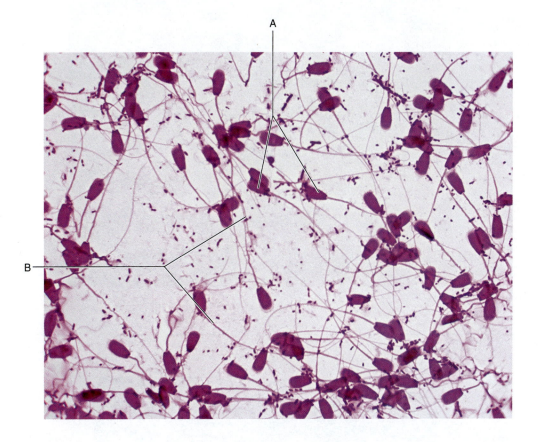

A. _____   B. _____

Exercise 11.12 **Testis, Cross Section 40x and 400x**

GO TO HISTOLOGY > REPRODUCTIVE SYSTEM > SELF REVIEW > IMAGES 4 AND 6

- *Mouse over the images to locate and label the structures indicated below. Click on the
  structures to hear their pronunciations.*

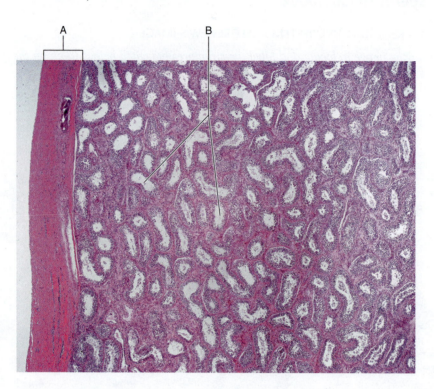

A. _____   B. _____

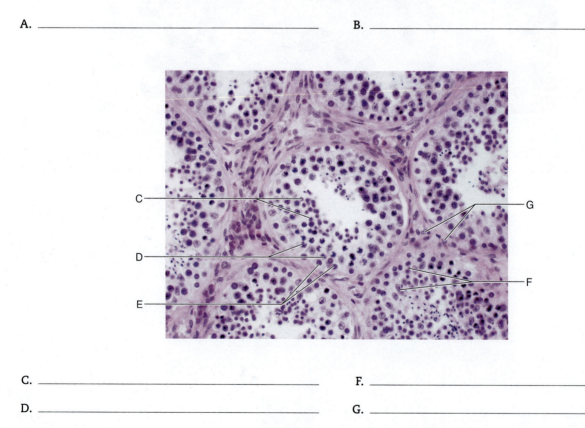

C. _____   F. _____

D. _____   G. _____

E. _____

Exercise 11.13 **Ovary, Sagittal Section 100x**

GO TO ⟩ HISTOLOGY > REPRODUCTIVE SYSTEM > SELF REVIEW > IMAGES 12 AND 13

- *Mouse over the images to locate and label the structures indicated below. Click on the structures to hear their pronunciations.*

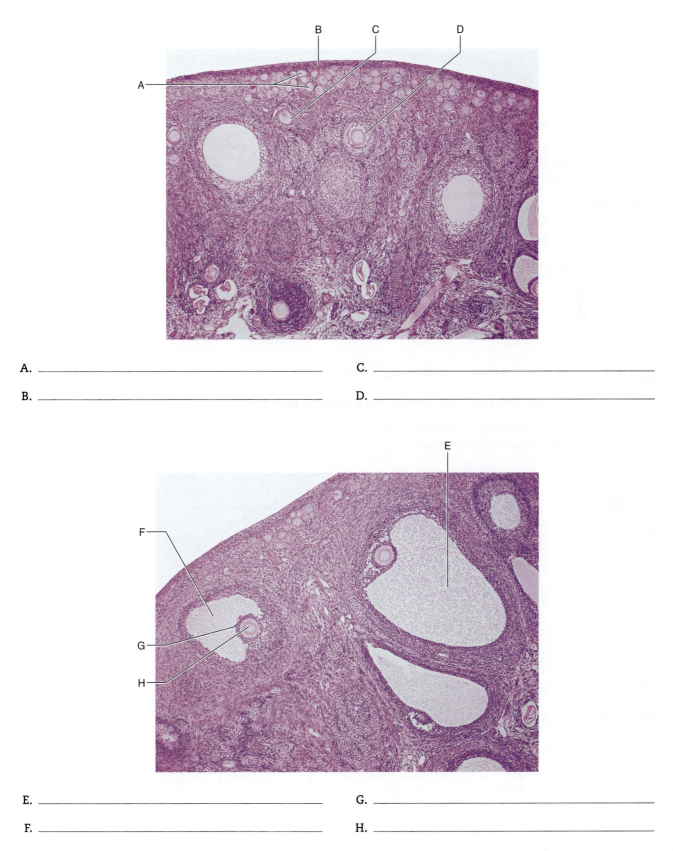

A. _____   C. _____

B. _____   D. _____

E. _____   G. _____

F. _____   H. _____

Exercise 11.14 Ovary, Corpus Luteum 40x

GO TO ⟩ HISTOLOGY > REPRODUCTIVE SYSTEM > SELF REVIEW > IMAGE 22

- *Mouse over the image to locate and label the structures indicated below. Click on the structures to hear their pronunciations.*

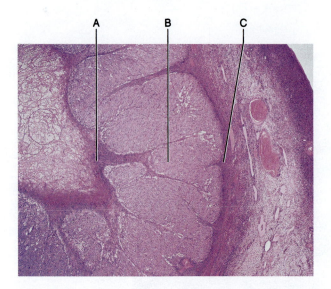

A. _____

B. _____

C. _____

Exercise 11.15 Ovary, Corpus Albicans 100x

GO TO ⟩ HISTOLOGY > REPRODUCTIVE SYSTEM > SELF REVIEW > IMAGE 27

- *Mouse over the image to locate and label the structure indicated below. Click on the structure to hear its pronunciation.*

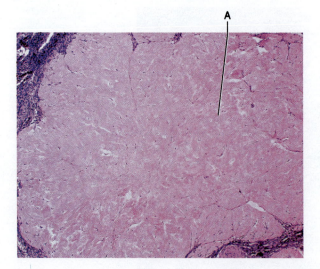

A. _____

Exercise 11.16   **Uterus, Secretory Phase, Cross Section 100x**

GO TO ⟩ HISTOLOGY > REPRODUCTIVE SYSTEM > SELF REVIEW > IMAGE 32

- *Mouse over the image to locate and label the structures indicated below. Click on the structures to hear their pronunciations.*

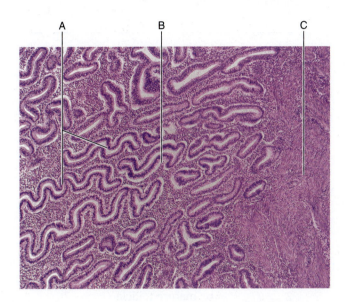

A. _____

B. _____

C. _____

Exercise 11.17   **Vagina, Cross Section 40x**

GO TO ⟩ HISTOLOGY > REPRODUCTIVE SYSTEM > SELF REVIEW > IMAGE 33

- *Mouse over the image to locate and label the structures indicated below. Click on the structures to hear their pronunciations.*

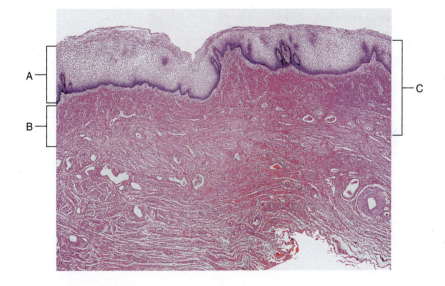

A. _____   C. _____

B. _____

QUIZ) Histology

Answer the following questions using information from PAL 3.1 as well as other course materials including your textbook, lecture, and lab notes.

I. Check Your Understanding

1. Match the structures, at right, with their correct location, at left.

   _____ capsule   a. spermatogonium

   _____ tail   b. mitochondria

   _____ basement   c. tunica albuginea
          membrane
                d. sperm

   _____ lumen   e. flagellum

   _____ midpiece

2. Match the structures, at right, with their correct location, at left.

   _____ endometrium   a. primordial follicle

   _____ corpus albicans   b. uterine glands

   _____ ovary   c. collagen fibers

   _____ vagina   d. ovum

   _____ corona radiata   e. stratified squamous
                        epithelium

3. Identify the glycoprotein-rich layer that surrounds an oocyte.

   _____

BEYOND PAL 4. Arrange the structures at right according to the events of spermatogenesis.

   1. _____   a. spermatid

   2. _____   b. spermatogonium

   3. _____   c. secondary spermatocyte

   4. _____   d. sperm

   5. _____   e. primary spermatocyte

5. Identify the cellular end product of spermatogenesis.

   _____

   _____

6. By which process does a spermatid become an anatomically mature sperm?

   _____

7. Which cells of the testis secrete androgens?

   _____

8. Arrange the structures at right according to the events of the ovarian cycle.

   1. _____   a. primary follicle

   2. _____   b. mature follicle

   3. _____   c. corpus luteum

   4. _____   d. corpus albicans

   5. _____   e. secondary follicle

   6. _____   f. primordial follicle

II. Apply What You Learned

Endometriosis is a common female reproductive disease. It may result from the backward flow of menstrual fluid, containing bits of endometrium, through the uterine tubes and out into the peritoneal cavity.

1. What is endometriosis?

_____

_____

_____

_____

_____

_____

_____

2. Why can endometriosis be a painful condition?

_____

_____

_____

_____

_____

_____

3. What is the major, long-term complication associated with endometriosis?

_____

_____

_____

_____

_____

LAB PRACTICAL   Reproductive System

1. Identify the structure.

   _____

2. Identify the structure.

   _____

3. Identify the structure.

   _____

4. Identify the structure.

   _____

5. Identify the structure.

   _____

6. Identify the structure.

   _____

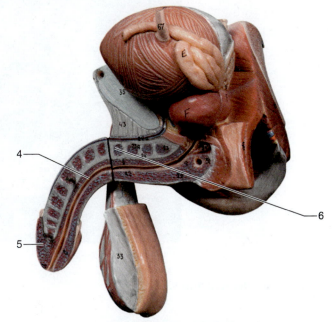

Copyright by SOMSO, 2010, www.somso.com

LAB PRACTICAL *continues*

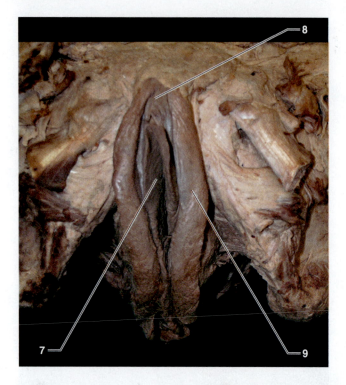

7. Identify the structures.

_____

8. Identify the structure.

_____

9. Identify the structures.

_____

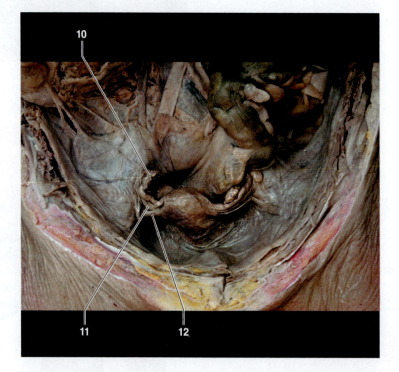

10. Identify the structures.

_____

11. Identify the structure.

_____

12. Identify the structure.

_____

13. Identify the structure.

_____

14. Identify the structure.

_____

15. Identify the structure.

_____

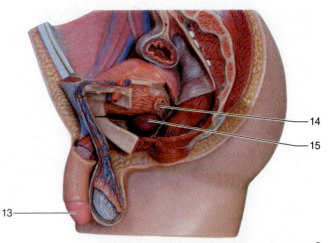

LT-H11: Male pelvis, 2-part, 3B Scientific®

16. Identify the structure.

_____

17. Identify the structure.

_____

18. Identify the structure.

_____

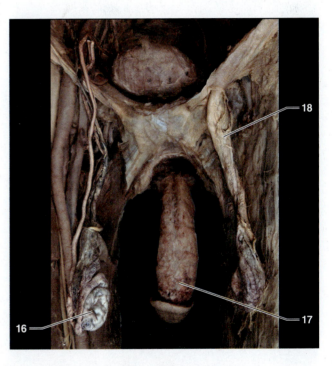

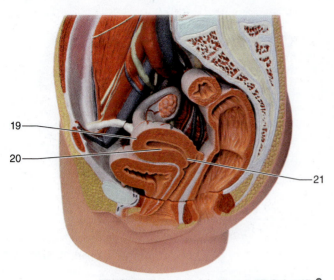

LT-VG366: Female pelvis, 2-part, 3B Scientific®

19. Identify the structure (be specific).

_____

20. Identify the structure (be specific).

_____

21. Identify the structure (be specific).

_____

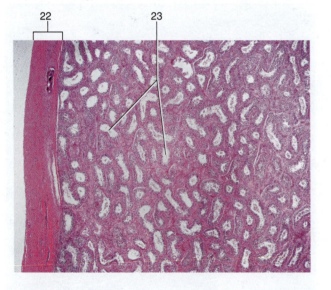

22. Identify the structure.

_____

23. Identify the structures.

_____

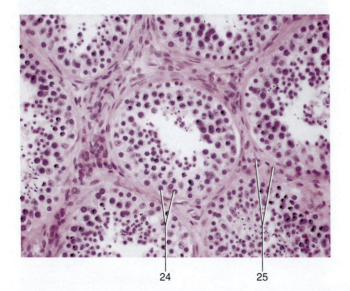

24. Identify the cells.

_____

25. Identify the cells.

_____

26. Identify the structures.

_____

27. Identify the structure.

_____

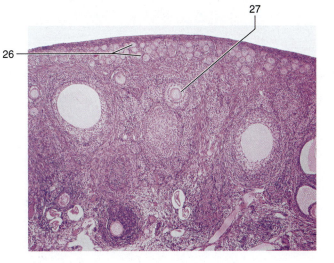

28. Identify the layer.

_____

29. Identify the layer.

_____

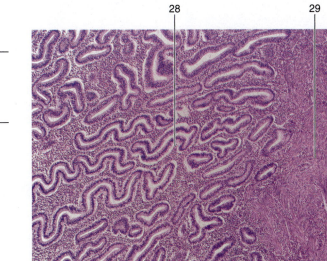

30. Identify the organ shown in the image at right.

_____

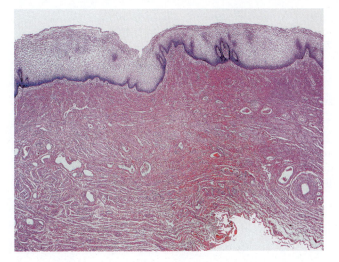